PROBABILISTIC NORMED SPACES

PROBABILISTIC NORMED SPACES

Bernardo Lafuerza Guillen
Panackal Harikrishnan

ICP Imperial College Press

Published by

Imperial College Press
57 Shelton Street
Covent Garden
London WC2H 9HE

Distributed by

World Scientific Publishing Co. Pte. Ltd.
5 Toh Tuck Link, Singapore 596224
USA office: 27 Warren Street, Suite 401-402, Hackensack, NJ 07601
UK office: 57 Shelton Street, Covent Garden, London WC2H 9HE

Library of Congress Cataloging-in-Publication Data
Lafuerza Guillén, Bernardo, author.
Probabilistic normed spaces / Bernardo Lafuerza Guillén, Universidad de Almeria, Spain, Panackal Harikrishnan, Manipal Institute of Technology (MIT Manipal), India.
pages cm
Includes bibliographical references and index.
ISBN 978-1-78326-468-1 (alk. paper)
1. Normed linear spaces. I. Harikrishnan, Panackal, author. II. Title.
QA322.2.L38 2014
515'.73--dc23
2014018361

British Library Cataloguing-in-Publication Data
A catalogue record for this book is available from the British Library.

Typeset by Stallion Press
Email: enquiries@stallionpress.com

Printed in Singapore

Preface

The theory of probabilistic normed spaces, henceforth called PN spaces, was born as a "natural" consequence of the theory of probabilistic metric spaces, henceforth PM spaces. These spaces were introduced by Karl Menger (Menger, 1942) who introduced the idea of a statistical metric, i.e. of replacing the number $d(p, q)$, which gives the distance between two points p and q in a non-empty set S, by a distribution function F_{pq} whose value $F_{p,q}(t)$ at $t \in]0, +\infty]$ is interpreted as the probability that the distance between the points p and q is smaller than t. The theory was then brought to its present state by Schweizer and Sklar in a series of papers (Schweizer & Sklar, 1958, 1960, 1962, 1963, 1973). The achievements of these two authors culminated in their masterly monograph (Schweizer & Sklar, 1983), in which all the known results until the early 1980s were collected. This book has recently been reprinted by Dover (Schweizer & Sklar, 2005). The reader should read the lucid introduction (Schweizer, 2003) in Menger's *Selecta* (Schweizer *et al.*, 2003), where Schweizer goes through the history of the subject. Thus PN spaces may be regarded as just a chapter, albeit an important one, in the theory of PM spaces.

PN spaces were introduced by Šerstnev in a series of papers (Šerstnev, 1962, 1963a, 1963b, 1964a); he was motivated by problems of best approximation in statistics. His definition runs along the same path followed in order to probabilize the notion of metric space and to introduce PM spaces. His definition turned out to be strict, perhaps too strict, so that after an initial flourishing of papers, mainly from the Russian school, the theory remained dormant for a number of years until it was revived by a new definition proposed in (Alsina, Schweizer, & Sklar, 1993), who reexamined in a profound way the definition of a classical norm, gave an equivalent formulation of a condition given by Šerstnev, and proposed the new definition. In this book we shall deal with PN spaces according to both definitions.

The first author and Carlo Sempi met in the early 1990s immediately after the drafting of the paper with the new definition and decided to start working on these new spaces. Many of the results presented here are the fruit of their cooperation on this subject. The first author and Sempi are thus grateful to PN spaces for bringing them together and for having caused the exchange of several visits between the University of Lecce and University of Almería. The idea of writing such a book came up during a Conference in Almería in June 2005, when Sempi, was invited to present a survey of the theory of PN spaces. That survey (Sempi, 2005) was the germ from which this book arose. The authors of this book met in June 2012 at the University of Almería for the scientific discussions regarding the recent developments in PN spaces, and finally that meeting made this book come into existence, with the moral support of Carlo Sempi.

Before beginning we must express our gratitude to Bert Schweizer and Abe Sklar, from whom we have learnt much of what we know, who have always been generous with their valuable advice and encouragement, and who offered us a unique opportunity to go through their papers before publishing them. In a sense, this book has the goal, perhaps too ambitious, to extend the results of their book to the setting of PN spaces. Similarly, in order to keep a correct sense of proportions, the reader ought to bear in mind that practically all the tools and many of the concepts we use in this book were created by them. The authors would also like to acknowledge the moral support of Prof. Carlo Sempi, University of Lecce, Italy for turning this book into a reality.

There is a long list of friends whom we wish to thank for their constant support, encouragement, valuable ideas, and ever-enduring friendship. José Juan Quesada Molina acted as a midwife, so to speak, because it was he who engineered our first getting together when he perceived that we could collaborate with each other to form a team. In fact, he assessed our different personalities and foresaw that we could form a competent team; for this, for his friendship over a long period of time, and for the many valuable discussions and thought-provoking conversations we had with him, we owe him many thanks. Claudi Alsina and Roger Nelsen have always been in our minds as potential partners as well as friends.

We extend our thanks to Dr K.T. Ravindran (Payyanur College, Kerala) for his valuable suggestions during the preparation of the content presented in Chapter 11 of the book. The authors would like to gratefully and sincerely thank Dr Vinod V. Thomas, Director, MIT, Manipal University and Dr Srikanth Rao, Joint Director, MIT, Manipal University for their kind support and guidance. The second author extends his thanks to Dr P.G. Bhat, HOD of Mathematics, MIT, Manipal University for his encouragement. The second author also remembers with gratitude the support and inspiration from his colleagues Dr Kuncham Syam Prasad, Dr Kedukkodi Babu Shri Srinivas, Dr Srikanth Prabhu, Dr Baiju T., Dr Vadiraja G. Bhatta, Mr. Mohana K.S., and Mrs. Prathima J. of MIT, Manipal University.

The first author expresses his heartfelt thanks to his wife, María del Mar, for her incessant support, without which he would never have been able to complete this challenging undertaking.

The second author expresses his deep sense of gratitude and appreciation to Anjali (his wife), Poornasree (his daughter), and others for their continued support, inspiration, encouragement, quiet patience, and unwavering love that made it possible for him to pursue higher studies in Mathematics.

Suggestions for the improvement of the book will be gratefully acknowledged.

Bernardo Lafuerza Guillén
Panackal Harikrishnan

Contents

Chapter 1

Preliminaries

It is impossible to speak about probabilistic normed spaces (PN spaces) without making reference to the concept of a probabilistic metric space; several definitions and preliminaries will be needed. Most of the concepts listed below are studied in depth in the fundamental book by Schweizer & Sklar (1983). In this chapter we collect all the concepts and the tools that will be necessary in order to study PN spaces and we fix the notation.

1.1 Probability Spaces

A probability space is a triple $(\Omega, \mathcal{F}, P)$[1] where Ω is a nonempty set and $\mathcal{F}$ is a σ-algebra of subsets of Ω, i.e. a family of subsets of Ω such that

(a) the set Ω belongs to $\mathcal{F}, \Omega \in \mathcal{F}$;
(b) the complement A^c of every set A in $\mathcal{F}$ also belongs to $\mathcal{F}$;
(c) the countable union of a sequence $\{A_n : n \in \mathbb{N}\}$ of subsets belonging to $\mathcal{F}$, $A_n \in \mathcal{F}$ for every $n \in \mathbb{N}$ belongs to $\mathcal{F}$, i.e. $\cup_{n \in \mathbb{N}} A_n \in \mathcal{F}$.

A *random variable* on it is a function $f : \Omega \to \overline{\mathbb{R}}$ that is measurable with respect to the σ-algebras $\mathcal{F}$ and $\mathcal{B}$, where $\mathcal{B}$ is the Borel σ-algebra in

[1] Then $L_1^+(\Omega)$ is the set of all functions f from Ω into $\mathbb{R}$ that are measurable with respect to $\mathcal{F}$ and satisfy $P\{\omega \in \Omega | f(\omega) < 0\} = 0$. Since a probability measure, like any function, determines its domain, it is always possible — and often convenient — to speak of a function as being P-measurable, rather than *measurable with respect to the sigma field* $\mathcal{F} = \text{Dom}P$. Similarly, any property that holds everywhere on Ω except possibly on a set of probability 0 is said to hold *almost everywhere (a.e.)* or *almost surely (a.s.)*. Thus the elements of $L_1^+(\Omega)$ are the P-measurable functions that are a.s. non-negative.

$\overline{\mathbb{R}}$, *viz.* for every Borel set $B \in \mathcal{B}$ its inverse image defined by

$$f^{-1}(B) := \{\omega \in \Omega : f(\omega) \in B\}$$

belongs to the σ-algebra $\mathcal{F}$.

This brief section is not the right place to delve into probability. However the reader may find in many books an excellent treatment of this subject; here we mention only Dudley (1989); Williams (1991); Bauer (1996); and Jacod & Protter (2000).

Notice that the concept of random variable here considered is more general than that commonly used; in fact we allow a random variable f to take the values $-\infty$ and $+\infty$ with non-zero probability.

1.2 Distribution Functions

A distribution function (d.f.) is a function F defined on the extended real line $\overline{\mathbb{R}} := [-\infty, +\infty]$ that is increasing[2] (i.e. $F(t) \leq F(t')$ whenever $t \leq t'$), left-continuous on the set of real numbers $\mathbb{R}$ and such that $F(-\infty) = 0$ and $F(+\infty) = 1$. The set of all d.f.s will be denoted by Δ; the subset of Δ formed by the *proper* d.f.s, i.e. by those d.f.s F for which

$$\lim_{n \to -\infty} F(t) = 0 \quad \text{and} \quad \lim_{n \to +\infty} F(t) = 1$$

will be denoted by $\mathcal{D}$. The space $\mathcal{D}$, which is the one encountered in most probability textbooks, corresponds to the so-called *proper* d.f.s; these are the d.f.s of the real random variables, i.e. of those random variables f that almost surely take real values, $P(|f| = +\infty) = 0$.

A special family of d.f.s that will be frequently used is formed by the d.f.s $\{\varepsilon_a\}$, $a \in \overline{\mathbb{R}}$ defined via

$$\varepsilon_a(t) := \begin{cases} 0, & \text{if } t \leq a \\ 1, & \text{if } t > a \end{cases}$$

when $a \in \mathbb{R}$, while $\varepsilon_\infty(t) = 0$ for $t \in \mathbb{R}$ and $\varepsilon_\infty(+\infty) = 1$.

A d.f. has a probability meaning. Given a probability space $(\Omega, \mathcal{F}, P)$ and a random variable f defined on it, the d.f. F of the random variable f is defined by

$$F(t) := P(f \leq t) = P(\{\omega \in \Omega : f(\omega) \leq t\}) \tag{1.2.1}$$

It is well known that for every d.f. F there is a probability space $(\Omega, \mathcal{F}, P)$ and a random variable f defined on it such that the d.f. f is F (see, e.g., Billingsley, 1979).

[2]Throughout this book the word *increasing* is always understood in the weak sense of *non-decreasing*.

A natural ordering is introduced in Δ by setting $F \leq G$ if $F(t) \leq G(t)$ for every $t \in \mathbb{R}$. The maximal element in this order is the d.f. $\varepsilon_{-\infty}$ while $\varepsilon_{+\infty}$ is the minimal element. Finally notice that for every $a \in \mathbb{R}$, the d.f. ε_a is in $\mathcal{D}$ such that $\varepsilon_a \leq \varepsilon_b$ if, and only if, $b \leq a$.

It should be noticed that, given a set of d.f.s, its supremum is always a d.f., whereas its infimum may well not be a d.f., because it need not be left continuous. In order to see this, it suffices to consider the sequence $\{\varepsilon_{1/n}\}$, whose infimum is the function defined by

$$\varphi(t) := \begin{cases} 0, & \text{if } t < 1 \\ 1, & \text{if } t \geq 1 \end{cases}$$

which is not left continuous at the single point $t = 1$.

From a probabilistic point of view, it is important to consider *weak convergence* of d.f.s; a sequence $\{F_n\}$ of d.f.s is said to converge weakly to a d.f. F if the sequence of real numbers $\{F_n(t)\}$ converges to $F(t)$ for every point t at which the limit d.f. F is continuous; if

$$C(F) := \{t \in \mathbb{R} : F(t) = F(t+0)\}$$

is the set of continuity points for F,

$$\forall\, t \in C(F) \qquad \lim_{n\to\infty} F_n(t) = F(t).$$

It is known that in the set $\mathcal{D}$ one can introduce a metric d_L, called the Levy metric, such that the topology of the metric space $(\mathcal{D}, d_L)$ is that of weak convergence. The metric d_L was introduced by Paul Levy (Levy, 1937). The metric space $(\mathcal{D}, d_L)$ is complete, but not compact. For the proofs of the statements we have just made the reader may refer to Lukacs (1975a). Given two d.f.s F and G in $\mathcal{D}$, and let $h \in (0, 1]$, the Levy distance d_L is defined as follows

$$d_L(F, G) := \inf\left\{h > 0 : \forall x \in \left(-\frac{1}{h}, \frac{1}{h}\right) \right. \\ \left. F(x-h) - h \leq G(x) \leq F(x+h) + h\right\}. \tag{1.2.2}$$

However the Levy distance d_L is not appropriate in the case of the space Δ. Several distances on Δ (Sibley, 1971; Schweizer, 1975; Sempi, 1982; Taylor, 1985) metrize the topology of weak convergence, but, following Schweizer and Sklar (Section 4.2 in Schweizer & Sklar, 1983) we shall adopt the Sibley metric d_S.[3]

[3]Notice that in Schweizer & Sklar (1983) the Sibley metric is called the generalized Levy metric.

The Sibley metric (Sibley, 1971) is defined as follows: if F and G are d.f.s and h is in $]0,1[$, let $(F,G;h)$ denote the condition

$$F(x-h)-h \leq G(x) \leq F(x+h)+h \quad \text{for all } x \in \left]-\frac{1}{h},\frac{1}{h}\right[. \qquad (1.2.3)$$

Then the Sibley metric is defined by

$$d_S(F,G) := \inf\{h \in]0,1[: \text{ both } (F,G;h) \text{ and } (G,F;h) \text{ hold}\}. \qquad (1.2.4)$$

Since every d.f. F is bounded, $0 \leq F(t) \leq 1$, for all $t \in \overline{\mathbb{R}}$, one has $d_S \leq 1$. Notice that the difference between the Levy and Sibley metrics lies in the interval to which the variable x belongs.

We shall summarize in the following theorem the relevant facts about the Sibley metric; for their proof the reader is referred to Section 4.2 in Schweizer & Sklar (1983).

Theorem 1.2.1.

(a) *The function* $d_S : \Delta \times \Delta \to \mathbb{R}_+ = [0,\infty]$ *defined by* (1.2.3) *and* (1.2.4) *is a metric on* Δ;
(b) *If* $d_S(F,G) = h > 0$, *then both* $(F,G;h)$ *and* $(G,H;h)$ *hold*;
(c) *The metric space* (Δ, d_S) *is complete and compact*;
(d) *The topology of the metric* d_S *and the topology of weak convergence coincide, i.e. the sequence* $\{F_n\}$ *of d.f.s converges weakly to the d.f.* F *if, and only if,*

$$d_S(F_n,F) \underset{n\to+\infty}{\longrightarrow} 0.$$

1.3 The Space of Distance of Distribution Functions

As will be seen, a special role will be played by the *distance* d.f.s, namely those d.f.s F such that $F(0)=0$; these form a subset of Δ denoted by Δ^+; the analogous subset of $\mathcal{D}$ is denoted by $\mathcal{D}^+$.

A d.f. ε_a belongs to Δ^+, and in fact to $\mathcal{D}^+$, if, and only if, a is positive, $a \geq 0$.

The restriction to Δ^+ of the order introduced on Δ has now ε_0 as the maximal element while the minimal element is still ε_∞.

We keep denoting by d_S the restriction to Δ^+ of the Sibley metric, since no possible confusion may arise. The following theorem summarizes some facts that will be useful in the sequel.

Theorem 1.3.1.

(a) (Δ^+, d_S) *is a closed subspace of* (Δ, d_S);
(b) *If the d.f.s* $F_n (n \in \mathbb{N})$ *and* F *belong to* Δ^+, *then* $\{F_n\}$ *converges weakly to* F *if, and only if,* $d_S(F_n, F) \underset{n \to +\infty}{\longrightarrow} 0$;
(c) (Δ^+, d_S) *is a closed, and hence compact and complete, subset of* (Δ, d_S).

In Δ^+ the Sibley metric d_S takes a simpler form, a fact that will be exploited several times. Since for any two d.f.s F and G in Δ^+, one has $F(0) = G(0) = 0$, so that whenever x is in the interval $]0, \frac{1}{h}[$ with $h \leq 1, x - h \leq 0$. Therefore, if $[F, G; h]$ denotes the condition

$$G(x) \leq F(x+h) + h \quad \text{for all } x \in \left]0, \frac{1}{h}\right[\tag{1.3.1}$$

then the Sibley metric equals, for all F and G in Δ^+

$$d_S(F, G) := \inf\{h \in]0,1[: \text{ both } [F, G; h] \text{ and } [G, F; h] \text{ hold.}\} \tag{1.3.2}$$

The following two lemmas are reproduced from Section 4.3 of the book by Schweizer and Sklar (1983) and are used frequently in calculations of the Sibley distance.

Lemma 1.3.1. *If* $d_S(F, G) = h > 0$ *for two d.f.s* F *and* G *in* Δ^+, *then both* $[F, G; h]$ *and* $[G, F; h]$ *hold.*

Lemma 1.3.2. *For every* $F \in \Delta^+$

$$\begin{aligned} d_S(F, \varepsilon_0) &= \inf\{h > 0 : [F, \varepsilon_0]\} \\ &= \inf\{h > 0 : F(h+0) > 1 - h\} \end{aligned} \tag{1.3.3}$$

and, for all $t > 0$,

$$F(t) > 1 - t \quad \textit{if, and only if,} \quad d_S(F, \varepsilon_0) < t. \tag{1.3.4}$$

The distance $d_S(F, \varepsilon_0)$ has the following geometric interpretation; it is the abscissa of the point of intersection of the straight line $y = 1 - x$ with the graph of F, completed, if need be, by the addition of vertical segments joining the points $(t_0, F(t_0))$ and $(t_0, F(t_0 + 0))$ if t_0 is a point of discontinuity for F. As a consequence one immediately has

Lemma 1.3.3. *If* F *and* G *are in* Δ^+ *and* $F \leq G$, *then* $d_S(F, \varepsilon_0) \geq d_S(G, \varepsilon_0)$.

1.4 Copulas

Definition 1.4.1. A copula is a function $C : [0,1]^2 \to [0,1]$ that satisfies the following conditions:

(C1) for every $t \in [0,1]$, $C(0,t) = C(t,0) = 0$ and $C(1,t) = C(t,1) = t$;
(C2) C is 2-increasing, i.e., for all s, s', t and t' in $[0,1]$, with $s \leq s'$ and $t \leq t'$,

$$C(s',t') - C(s',t) - C(s,t') + C(s,t) \geq 0. \tag{1.4.1}$$

It follows from Definition 1.4.1 that every copula C is increasing in each place.

Moreover for any copula C one has $W \leq C \leq M$.

Definition 1.4.2. Let φ be a continuous, strictly decreasing function from $\mathbb{I} = [0,1]$ to $[0,+\infty]$ such that $\varphi(1) = 0$. The pseudo-inverse of φ is the function $\varphi^{[-1]}$ with Dom $\varphi^{[-1]} = [0,+\infty]$ and Ran $\varphi^{[-1]} = \mathbb{I}$ defined by

$$\varphi^{[-1]}(t); = \begin{cases} \varphi^{-1}(t), & 0 \leq t \leq \varphi(0), \\ 0, & \varphi(0) \leq t \leq +\infty. \end{cases} \tag{1.4.2}$$

The following results are well known and can be found in (Nelsen, 2006) or in (Schweizer & Sklar, 2005).

Lemma 1.4.1. *Let φ be a continuous, strictly decreasing function from $\mathbb{I}$ to $[0,+\infty]$ such that $\varphi(1) = 0$, and let $\varphi^{[-1]}$ be the pseudo-inverse of φ defined by* (1.4.2). *Let C be the function from $\mathbb{I}^2$ to $\mathbb{I}$ defined by*

$$C(u,v) = \varphi^{[-1]}(\varphi(u) + \varphi(v)). \tag{1.4.3}$$

Then C satisfies the boundary conditions (C1). *Moreover C is 2-increasing if, and only if, for all $v \in \mathbb{I}$,*

$$C(u_2,v) - C(u_1,v) \leq u_2 - u_1, \tag{1.4.4}$$

whenever $u_1 \leq u_2$.

Theorem 1.4.1. *Let φ be a continuous, strictly decreasing function from $\mathbb{I}$ to $[0,+\infty]$ such that $\varphi(1) = 0$, and let $\varphi^{[-1]}$ be the pseudo-inverse of φ defined by* (1.4.2). *Then the function C from $\mathbb{I}^2$ to $\mathbb{I}$ defined by* (1.4.3) *is a copula if, and only if, φ is convex.*

Copulas of the form (1.4.2) are called Archimedean. An Archimedean copula is associative and a t-norm.

1.5 Triangular Norms

Definition 1.5.1. A triangular norm (t-norm) T is a binary operation on the unit interval $T : [0,1] \times [0,1] \to [0,1]$ that is associative, commutative, increasing in each place, and which has 1 as identity, namely, for all x, y, and z in $[0,1]$, one has

(T1) $T(T(x,y),z) = T(x,T(y,z))$;
(T2) $T(x,y) = T(y,x)$;
(T3) $T \mapsto T(t,x)$ and $t \mapsto T(x,t)$ are increasing;
(T4) $T(1,x) = T(x,1) = x$.

The introduction of t-norms was motivated by the need to generalize the triangle inequality from the setting of metric spaces to that of probabilistic metric spaces (see below); then associativity was required in order to have an unambiguous extension to polygonal inequalities. The important monograph by Klemente, Mesiar, and Pap (2000) is entirely devoted to t-norm; but see the survey by Klemente and Mesiar (2005). The most general solution of the system (T1)–(T4) is not known. However, the problem of finding continuous t-norms is equivalent to the problem of finding certain continuous solutions to the functional equations of associativity or, in a different terminology, certain topological semigroups, on the unit interval. This problem is solved: its well-known solution — the representation of an arbitrary topological semigroup on $[0,1]$, with unit 1 and null-element 0, as an ordinal sum of three non-isomorphic types — yields large numbers of t-norms. A t-norm T is *strict* if it is continuous on $[0,1] \times [0,1]$ and it is strictly increasing on $[0,1] \times [0,1]$. It is *Archimedean* if it has no idempotents other than 0 and 1, i.e. if $T(x,x) < x$ for every $]0,1[$. For Archimedean copulas see the books by Schweizer and Sklar (1983), Nelsen (2006), and Klemente, Mesiar, and Pap (2000).

Examples of t-norms are M, W, Π, and Z are defined respectively by

$$M(x,y) := min\{x,y\} = x \wedge y$$

$$W(x,y) := max\{x+y-1, 0\} = 0 \vee (x+y-1)$$

$$\Pi(x,y) := xy$$

$$Z(x,y) := \begin{cases} 0, & \text{if } (x,y) \in [0,1[\times[0,1] \\ x, & \text{if } x \in [0,1], y = 1 \\ y, & \text{if } x = 1, y \in [0,1]. \end{cases}$$

One has

$$Z < W < \Pi < M,$$

and, for every t-norm T,

$$Z \leq T \leq M.$$

Together with a t-norm T, it is often important to consider its *t-conorm* T^* defined by

$$T^*(x, y) := 1 - T(1 - x, 1 - y).$$

A t-conorm T^* satisfies properties (T1), (T2), and (T3) of Definition 1.5.1 and

$$\forall x \in [0, 1] T^*(x, 0) = T(0, x) = x. \tag{1.5.1}$$

A function $S : [0, 1] \times [0, 1] \to [0, 1]$ is said to be *s-norm* if it is associative, commutative, increasing in each place, and has identity, in other words, if it satisfies properties (T1), (T2), and (T3) of Definition 1.5.1 and the boundary condition (1.5.1). Obviously, for every s-norm S there is a t-norm T of which S is the t-conorm:

$$S(x, y) = T^*(x, y).$$

There are good reasons to treat s-norms independently of t-conorms (Schweizer, 2005). A t-norm T is a copula if, and only if, it satisfies the Lipschitz condition $T(c, b) - T(a, b) \leq c - a$ for all $a, b, c \in [0, 1]$ with $a \leq c$.

1.6 Triangle Functions

Definition 1.6.1. A triangle function τ is a binary operation on Δ^+, namely a mapping $\tau : \Delta^+ \times \Delta^+ \to \Delta^+$ that is associative, commutative, increasing in each place, and which has ε_0 as unit, *viz.* for all F, G, and H in Δ^+; explicitly, a triangle function τ satisfies the following conditions, for all F, G, and H in Δ^+;

(TF1) $\tau(\tau(F, G), H) = \tau(F, \tau(G, H))$;
(TF2) $\tau(F, G) = \tau(G, H)$;
(TF3) if $F \leq G$. then both $\tau(F, H) \leq \tau(G, H)$ and $\tau(H, F) \leq \tau(H, G)$;
(TF4) $\tau(\varepsilon_0, F) = \tau(F, \varepsilon_0) = F$.

Triangle functions can be constructed through left-continuous t-norms: if T is such a t-norm then

$$\tau_T(F, G)(x) := \sup\{T(F(s), G(t)); s + t = x\} \tag{1.6.1}$$

is a triangle function. If, moreover, T is continuous, then τ_T is uniformly continuous on (Δ^+, d_S).

But triangle functions can be constructed through t-conorms: if T^* is a continuous t-conorm, then

$$\tau_{T^*}(F,G)(x) := \ell^- \inf\{T^*(F(s),G(t)); s+t=x\}$$

is a triangle function, which is uniformly continuous on (Δ^+, d_S). Here $\ell^-\varphi(x)$ represents the left limit at x, $\ell^-\varphi(x) = \varphi(x-0) := \lim_{t\to x, t<x} \varphi(t)$.

It is important to notice that one has, for all a and b in $\mathbb{R}_+$,

$$\tau_T(\varepsilon_a, \varepsilon_b) = \varepsilon_{a+b} = \tau_{T^*}(\varepsilon_a, \varepsilon_b). \tag{1.6.2}$$

Other notable examples of triangle functions are provided, for every pair F and G of d.f.s in Δ^+

$$(F * G)(x) := \int_{[0,1[} F(x-t)dG(t), \tag{1.6.3}$$

(see Wald, 1943), and, for every continuous t-norm T, by

$$\Pi_T(F,G)(x) := T(F(x),G(x)). \tag{1.6.4}$$

Notice that Π_{T^*} is not a triangle function; indeed, although it is associative, commutative, and increasing in each place, ε_0 is not its unit, but rather its annihilator: for all $x \in [0,\infty]$,

$$\begin{aligned} \Pi_{T^*}(\varepsilon_0, F)(x) &= T^*(\varepsilon_0(x), F(x)) = 1 - T(1-\varepsilon_0(x), 1-F(x)) \\ &= 1 - T(0, \overline{F}(x)) = 1 = \varepsilon_0(x). \end{aligned}$$

Thus the relationship between Π_{T^*} and Π_T is similar to that between a t-norm and its t-conorm.

A notable binary operation on Δ^+ is defined, for all d.f.s F and G in Δ^+, by

$$\sigma_C(F,G)(t) := \int_{\{(u,v)\in\mathbb{R}_+; u+v\leq t} dC(F(u),G(v)), \tag{1.6.5}$$

where C is a copula; it is to be noted that not for every copula C is σ_C a triangle function because σ_C need not be associative. The following result was proved in Frank (1975).

Theorem 1.6.1. *For every copula C (1.6.5) defines a binary operation on Δ. This is associative on Δ, or on Δ^+, and hence, in this latter case, a triangle function, if, and only if, one of the following conditions hold:*

(a) $C = M$;
(b) $C = \Pi$;

(c) $([0,1], C)$ *is an ordered sum of copulas, each of which is either* M *or* Π.

As will be seen in the following it is often of interest to know when a triangle function τ is a binary operation on the set $\mathcal{D}^+$ of proper distance d.f.s, or in other words, when $\mathcal{D}^+$ is *stable* under τ, namely, when

$$\tau(\mathcal{D}^+ \times \mathcal{D}^+) \subset \mathcal{D}^+. \tag{1.6.6}$$

A sufficient condition for (1.6.3) is provided by the following theorem.

Theorem 1.6.2. *If one of the following conditions holds*:

(a) $\tau = \Pi_T$ *for some left-continuous t-norm* T;
(b) $\tau = \tau_T$ *for some continuous t-norm* T;
(c) $\tau = \sigma_C$ *for some copula* C;
(d) τ *is the convolution* $*$,

then the set $\mathcal{D}^+$ *is stable under* τ. *i.e.* $\tau(\mathcal{D}^+ \times \mathcal{D}^+) \subset \mathcal{D}^+$.

Proof. Let F and G be d.f.s belonging to $\mathcal{D}^+$, i.e. such that

$$\lim_{t\to+\infty} F(t) = \lim_{t\to+\infty} G(t) = 1.$$

(a) If $\tau = \Pi_T$ for some left-continuous t-norm T, then

$$\lim_{t\to+\infty} \Pi_T(F,G)(t) = T(F(t), G(t)) = T(1,1) = 1,$$

so that also $\Pi_T(F,G)$ belongs to $\mathcal{D}^+$.

(b) In view of the definition of $\tau_T(F,G)$, one has, for every $x > 0$,

$$\tau_T(F,G)(x) \geq T(F(x/2), G(x/2)).$$

Now, let x tend to $+\infty$ in order to obtain

$$\ell^- \tau_T(F,G)(+\infty) \geq T(1,1) = 1.$$

Thus $\tau_T(F,G)$ is in $\mathcal{D}^+$.

(c) It is known (see Theorem 4 in Schweizer and Sklar (1974)) that if X and Y are positive real-valued random variables on a probability space $(\Omega, \mathcal{A}, P)$, having continuous d.f.s F_X and F_Y, if C is their (unique) copula, then the d.f. F_{X+Y} of their sum $X+Y$ is given by

$$F_{X+Y}(t) = \int_{\{(u,v)\in\mathbb{R}_+ : u+v\leq t\}} dC(F_X(u), F_Y(v)).$$

But, since both X and Y are real-valued, both

$$P(X < +\infty) = 1 \quad \text{and} \quad P(Y < +\infty) = 1,$$

or, equivalently,

$$\lim_{t \to +\infty} F_X(t) = 1 \quad \text{and} \quad \lim_{t \to +\infty} F_Y(t) = 1$$

hold. As a consequence, their sum $X+Y$ is also a.c. finite. $P(X+Y < +\infty) = 1$, or, equivalently

$$1 = \lim_{t \to +\infty} F_{X+Y}(t) = \lim_{t \to +\infty} \int_{\{(u,v) \in \mathbb{R}_+ : u+v \leq t\}} dC(F_X(u), F_Y(v)),$$

which proves the assertion. More directly $\ell^- \sigma_C(+\infty)$ represents the probability that the random vector (X, Y) takes value in the first quadrant $[0, +\infty[\times[0, +\infty[$, and this is equal to 1.

(d) This is particular case of the previous one, when $C = \Pi$, or equivalently, when the two (continuous) random variables X and Y are independent. □

1.7 Multiplications

A multiplication is a binary operation on Δ^+; it generalizes the notion of triangle function. The results we present can be found in Schweizer (1975). Various investigations in probability theory and related areas lead to the problem of finding topological semigroups on spaces of d.f.s. One semigroup, convolution, has been studied in great detail. Several others, e.g., generalized convolutions and Delphic semigroups have received some attention. The general problem is unsolved.

1.7.1 *The multiplication τ_T*

Subject to mild restrictions on T, each such τ_T given by (1.6.1) is a triangle function. This result is due to Šerstnev (1963).

Theorem 1.7.1. *If T is a left-continuous t-norm and $F, G \in \Delta^+$ then the function $\tau_T(F, G)$ defined on $\mathbb{R}$ by* (1.6.1) *belongs to Δ^+.*

Theorem 1.7.2. *If T is a continuous t-norm and if one of the functions F, G is continuous, then the function $\tau_T(F, G)$ defined by* (1.6.1) *is continuous.*

Theorem 1.7.3. *If T is a left-continuous t-norm then the function τ_T given by* (1.6.1) *is also associative, hence it is a triangle function.*

Theorem 1.7.4. *If T is a continuous t-norm then τ_T is continuous (as a two-place function) on the metric space (Δ^+, d_S).*

Corollary 1.7.1. *Under the hypotheses of Theorem 1.7.4, τ_T is uniformly continuous on (Δ^+, d_S).*

1.7.2 *The multiplication Π_T*

One considers another class of triangle functions, also induced by t-norms, that arise naturally in the study of probabilistic metric spaces and play an important role in the probabilistic version of the Kampé de Feriét–Forte theory of information. They have not been employed in the triangle inequality $F_{pr} \geq \tau(F_{pq}, F_{qr})$ itself because the geometric restrictions they impose are generally too strong.

Theorem 1.7.5. *Let T be a left-continuous t-norm and let Π_T be the function defined for any $F, G \in \Delta^+$ and any $x \in \mathbb{R}$ by*

$$\Pi_T(F, G)(x) = T(F(x), G(x)). \tag{1.7.1}$$

Then Π_T is a triangle function.

Theorem 1.7.6. *If T is a continuous t-norm then the function Π_T from $\Delta^+ \times \Delta^+$ into Δ^+ defined by (1.7.1) is uniformly continuous (as a two-place function) on the metric space (Δ^+, d_S).*

1.7.3 *Convolution*

For any $F, G \in \Delta^+$ the convolution $F * G$, of F and G, is the function on $\mathbf{R}$ defined by (1.6.3). It is known that $F * G$ is in Δ^+ and further, since $F(-\infty) = G(-\infty) = 0$, that convolution in Δ^+ satisfies the properties (TF1) to (TF4) from Definition (1.6.1). We shall state all this as a Theorem.

Theorem 1.7.7. *Convolution is a triangle function.*

To complete the picture, we give:

Theorem 1.7.8. *Convolution is uniformly continuous on (Δ^+, d_S).*

1.7.4 *Convolution-related operations and random variables*

Convolution is a single binary operation whereas the operations τ_T and Π_T are families of such operations — one for each left-continuous t-norm T.

A. Sklar has noted that convolution may also be embedded in a family of binary operations on Δ^+ given by (1.6.4).

In addition for any copula C, we have

$$\tau_W \leq \tau_C \leq \sigma_C \leq \Pi_C \leq \Pi_{Min},$$

where τ_C and Π_C are the binary operations on Δ^+ defined by (1.6.1) and (1.6.4), respectively, with T replaced by C.

A t-norm T is a copula if, and only if, in addition to (T1)–(T4) of Definition (1.5.1), it satisfies (1.4.1); and a copula is a t-norm if, and only if, it is associative. Since a copula C may be non-associative, it is clear that not all functions σ_C are triangle functions. One might conjecture that σ_C is associative if, and only if, C is associative, but this is false. M. J. Frank has recently proved this remarkable result (see Theorem 1.6.1).

1.8 Probabilistic Metric Spaces

After the introduction of triangle functions, we are in a position to define probabilistic metric spaces. This theory started with Karl Menger (1942) who introduced the idea of a *statistical metric,* i.e. of replacing the number $d(p,q)$, which gives the distance between two points p and q, by a distribution function $F_{p,q}$ whose value $F_{p,q}(t)$ at $t \in]0,+\infty]$ is interpreted as the probability that the distance between the points p and q is smaller than t. The theory was then brought to its present state by Schweizer and Sklar in a series of papers (Schweizer & Sklar, 1958, 1960, 1962, 1963, 1973). The reader should read the lucid and masterly introduction (Schweizer, 2003) in Menger's *Selecta* (Schweizer *et al.*, 2003), where the author goes through the history of the subject. For a condensed presentation of PM spaces see also Sempi (2004).

Definition 1.8.1.
A *probabilistic metric space* (briefly, a PM space) is a triple $(S, \mathcal{F}, \tau)$, where S is a non-empty set, $\mathcal{F}$ is a function $S \times S$ into Δ_+ and τ is a triangle function, and the following conditions are satisfied for all points p, q, and r in S

(M1) $\mathcal{F}(p,p) = \varepsilon_0$;
(M2) $\mathcal{F}(p,p) \neq \varepsilon_0$ if $p \neq q$;
(M3) $\mathcal{F}(p,q) = \mathcal{F}(q,p)$;
(M4) $\mathcal{F}(p,r) \geq \tau(\mathcal{F}(p,q), \mathcal{F}(q,r))$.

If (M2) is not satisfied then one speaks of a *probabilistic Pseudo Metric space* (PPM space). If $(S, \mathcal{F}, \tau)$ is a PM space, one also says that $(S, \mathcal{F})$ is a PM space under τ. If $\tau = \tau_T$ for some t-norm T, then $(S, \mathcal{F}, \tau_T)$ is called a *Menger space under* T. If τ is the convolution then $(S, \mathcal{F})$ is said to be a *Wald space.* Usually one writes F_{pq} other than $\mathcal{F}(p, q)$.

Notice that, if (S, d) is a metric space, then $(S, \mathcal{F}, \tau)$, where the mapping $\mathcal{F} : S \times S \to \Delta^+$ is defined by

$$F_{pq} := \varepsilon_{d(p,q)} \tag{1.8.1}$$

and τ is a triangle function that, for all a and b in $\mathbb{R}_+$, satisfies (1.6.1) is a PM space. Conversely, for every PM space $(S, \mathcal{F}, \tau)$, in which the triangle function τ satisfies (1.6.1), there is a mapping $d : S \times S \to \mathbb{R}_+$ such that (1.8.1) holds; in this case the semigroups $(\mathbb{R}_+, +)$ and $(\{\varepsilon_a : a \in \mathbb{R}_+\}, \tau)$ are isomorphic. Therefore, the family of metric spaces may be regarded as a (proper) subset of the class of all PM spaces.

1.9 L^p and Orlicz Spaces

Since PN spaces have applications for L^p and *Orlicz space*, we devote this section to a very brief introduction to these latter spaces, which usually do not belong to the background of most mathematicians. For a thorough introduction to Orlicz spaces we refer to the books by M.S. Krasnosel'skiĭ and Y.B. Rutickiĭ (1961) and Rao & Ren (1991), where all the proofs of the results given in this section can be found.

We begin with the definition of a *Young function.* To this purpose we need the following notation.

Let $p : \mathbb{R}_+ \to \mathbb{R}_+$ be an increasing, right-continuous function such that

$$p(0) = 0, \quad \forall\, t > 0\ p(t) > 0 \quad \lim_{t \to +\infty} = +\infty. \tag{1.9.1}$$

Let $q : \mathbb{R}_+ \to \mathbb{R}_+$ be a right-continuous quasi-inverse of p,

$$q(t) := \sup\{s : q(s) \leq t\}.$$

The function q satisfies the conditions (1.9.1) and is increasing and right-continuous.

Definition 1.9.1. The functions φ and ψ from $\mathbb{R}_+$ into $\overline{\mathbb{R}}_+$ defined by

$$\varphi(x) := \int_0 p(t)dt \quad \text{and} \quad \psi(x) := \int_0 q(t)dt, \tag{1.9.2}$$

are called *complementary Young functions.* Each of the two functions φ and ψ is called a *Young function.*

Young functions are obviously symmetric and convex. In the following, we shall denote by $\mathcal{Y}$ the set of all Young functions.

Definition 1.9.2. Let $(\Omega, \mathcal{F}, \mu)$ be a finite atomless measure space, $\mu(\Omega) < +\infty$. Given a Young function ϕ, define for a measurable real function $f, f \in \mathcal{L}^0(\mathcal{F})$

$$\rho(f, \varphi) := \int \varphi \circ f d\mu.$$

Then the *Orlicz class* $C_\varphi(\Omega, \mathcal{F}, \mu)$ is defined by

$$C_\varphi(\Omega, \mathcal{F}, \mu) := \{f \in \mathcal{L}^0(\mathcal{F}); \rho(f, \varphi) < +\infty\}.$$

An Orlicz class C_φ satisfies the following inclusions

$$\mathcal{L}^\infty \subset C_\varphi \subset \mathcal{L}^1.$$

Both inclusions are strict. However, one has the following result:

Theorem 1.9.1. $\mathcal{L}^1 = \cup\{C_\varphi : \varphi \in \mathcal{Y}\}$.

For the comparison of the Orlicz classes generated by two different Young functions the following result is useful.

Theorem 1.9.2. *If φ_1 and φ_2 are two Young functions, the following statements are equivalent*

(a) $C_{\varphi_1} \subset C_{\varphi_2}$;
(b) *there exist x_0 and k in $]0, +\infty[$ such that, for all $x \geq x_0$,*

$$\varphi_2(x) \leq k\varphi_1(x).$$

As a consequence, two Orlicz classes C_{φ_1} and C_{φ_2} are equal if, and only if, there exist x_0, k, and k' in $]0, +\infty[$ such that, for all $x \geq x_0$,

$$k'\varphi_2(x) \leq \varphi_1(x) \leq k\varphi_2(x).$$

Among the Young functions those that satisfy an additional condition are particularly important.

Definition 1.9.3. The Young function φ is said to be *moderated* or to satisfy the Δ_2 condition, or again to be *moderate growth*, if there exist $k > 0$ and $x_0 \geq 0$ such that, for all $x \geq x_0$,

$$\varphi(2x) \leq k\varphi(x). \tag{1.9.3}$$

Notice that the function $|x| \mapsto |x|^p$, with $p > 1$ is moderated.

Any Orlicz class C_φ is convex, as is proved; however C_φ need not be a linear space. The following theorem highlights the importance of moderated Young functions.

Theorem 1.9.3. *For the Orlicz class C_φ the following statements are equivalent:*

(a) *C_φ is a vector space;*
(b) *the Young function φ is moderated.*

From now on, we shall assume that the spaces $\mathcal{L}^0$ and $\mathcal{L}^p, p \in [1, +\infty]$ have been replaced by their quotient with respect to the equivalence relationship $\simeq_\mu$ of being equal with respect to the measure μ; thus in effect, equivalence classes of functions, rather than individual functions, will be considered. As usual, we shall write

$$L^0 = \mathcal{L}^0|_{\simeq_\mu} \quad \text{and} \quad L^p = \mathcal{L}^p|_{\simeq_\mu}.$$

Given a pair of complementary Young functions φ and ψ, we shall put

$$L_\varphi = L_\varphi(\Omega, \mathcal{F}, \mu) := \left\{ f \in L^0 : \forall g \in C_\varphi |\langle\langle f, g\rangle| := \int fg d\mu < +\infty \right\}.$$

The following theorem states some of the important properties of L_φ.

Theorem 1.9.4. *Let φ and ψ be a pair of complementary Young functions; then*

(a) *L_φ is a vector space, called the Orlicz space, generated by φ;*
(b) *$C_\varphi \subset L_\varphi$;*
(c) *if f and g belong to C_φ and to C_φ, respectively, then the product fg is in L^1.*

The Orlicz space L_φ is a normed space.

Theorem 1.9.5. *Let φ and ψ be a pair of complementary Young functions; then, if f is in L_φ,*

$$\sup\{|\langle f, g\rangle| : g \in C_\varphi, \rho(g, \psi) \leq 1\} < +\infty. \tag{1.9.4}$$

The mapping $\|\cdot\|_\varphi : L_\varphi \to \mathbb{R}_+$ defined by

$$\|f\|_\varphi := \sup\{|\langle f, g\rangle| : g \in C_\varphi, \rho(g, \psi) \leq 1\}$$

is a norm on L_φ.

The case of the function $\varphi(x) := |x|^p/p, p > 1$ is of special relevance. In this case the complementary Young function ψ is given by $\psi(x) = |q|^q/q$ where q is the index conjugated to p, *viz.*

$$\frac{1}{p} + \frac{1}{q} = 1.$$

It turns out that, for such a φ

$$\|f\|_\varphi = q^{1/q}\|f\|_p,$$

where $\|\cdot\|_p$ is the usual norm in the space L^p. Therefore L^p spaces can be regarded as particular Orlicz spaces.

Theorem 1.9.6. *Every Orlicz space is complete, viz. it is a Banach space.*

In an Orlicz space it is possible to define another norm that is equivalent to the one that has just been introduced. This second norm is better suited for the considerations to be made regarding PN spaces.

Theorem 1.9.7. *Given a Young function φ the application $N_\varphi : L_\varphi \to \mathbb{R}_+$ defined by*

$$N_\varphi(f) := \inf\{k > 0 : \rho(f/k, \varphi) = \int \varphi\left(\frac{f}{k}\right) d\mu \le 1\} \tag{1.9.5}$$

is a norm on L_φ, called the Luxemburg norm. For every f in L_φ one has

$$N_\varphi(f) \le \|f\|_\varphi \le 2N_\varphi(f), \tag{1.9.6}$$

so that the two norms $N_\varphi(f)$ and $\|\cdot\|_\varphi$ generate the same topology.

1.10 Domination

The concept of domination in a partially ordered set will be needed when studying the product of PN spaces (Schweizer & Sklar, 1983; Sherwood, 1984; Tardiff, 1984).

Definition 1.10.1. Let $(S, \le)$ be a partially ordered set and let f and g be commutative and associative binary operations on S with a common identity e. Then f is said to *dominate* g, and one writes $f \gg g$, if, for all x_1, x_2, y_1, and y_2 in S, one has

$$f(g(x_1, y_1), g(x_2, y_2)) \ge g(f(x_1, x_2), f(y_1, y_2)). \tag{1.10.1}$$

Setting $y_1 = x_2 = e$ in (1.10.1) yields $f(x_1, y_2) \geq g(x_1, y_2)$, i.e., $f \geq g$. This implies that the dominance relationship is antisymmetric; it is also reflexive, but in general it need not be transitive. Apply this relationship to t-norms and triangle functions, in order to obtain:

Lemma 1.10.1. *The following statements hold*:

(a) *for every t-norm* $T, M \gg T$;
(b) *for every t-conorm* $T^*, T^* \gg M^*$;
(c) *for every continuous t-norm* $T, \Pi_T \gg \tau_T$;
(d) *for every triangle function* $\tau, \Pi_M \gg \tau$.

Proof. (a) For all x_1, x_2, y_1, and y_2 in $[0, 1]$, one has

$$M(x_1, x_2) \leq x_1 \quad \text{and} \quad M(y_1, y_2) \leq y_1;$$

thus

$$T(x_1, y_1) \geq T(M(x_1, x_2), M(y_1, y_2)).$$

Similarly

$$T(x_2, y_2) \geq T(M(x_1, x_2), M(y_1, y_2)),$$

whence

$$M(T(x_1, y_1), T(x_2, y_2)) \geq T(M(x_1, x_2), M(y_1, y_2)),$$

so that (1.10.1) holds with $f = M$ and $g = T$.

(b) Recall that $M^*(a, b) = max\{a, b\}$. Since $x_1 \leq M^*(x_1, y_1)$ and $x_2 \leq M^*(x_2, y_2)$.

$$T^*(x_1, y_1) \leq T^*(M^*(x_1, x_2), M^*(y_1, y_2)).$$

Similarly

$$T^*(x_2, y_2) \leq T^*(M^*(x_1, x_2), M^*(y_1, y_2)).$$

From the last two inequalities it follows that

$$T^*(M^*(x_1, x_2), M^*(y_1, y_2)) \geq M^*(T^*(x_1, y_1), T^*(x_2, y_2)),$$

namely $T^* \gg M^*$.

The proofs of (c) and (d) are very similar to that of (b); we give them only for the sake of completeness.

(c) For all d.f.s F_1, F_2, G_1, and G_2 in Δ^+ one has

$$\Pi_T(F_1, F_2) \leq \Pi_T(F_1, \varepsilon_0) \quad \text{and} \quad \Pi_T(F_1, F_2) \leq F_2,$$
$$\Pi_T(G_1, G_2) \leq \Pi_T(G_1, \varepsilon_0) \quad \text{and} \quad \Pi_T(G_1, G_2) \leq G_2.$$

Thus

$$\tau_T(\Pi_T(F_1,F_2),\Pi_T(G_1,G_2)) \le \tau_T(F_1,G_1),$$

$$\tau_T(\Pi_T(F_1,F_2),\Pi_T(G_1,G_2)) \le \tau_T(F_2,G_2),$$

which together yield, because Π_T is increasing in each place,

$$\Pi_T(\tau_T(F_1,G_1),\tau_T(F_2,G_2)) \ge \tau_T(\Pi_T(F_1,F_2),\Pi_T(G_1,G_2)),$$

namely the assertion.

(d) For all d.f.s F_1, F_2, G_1, and G_2 in Δ^+ one has

$$\Pi_M(F_1,F_2) \le F_1 \quad \text{and} \quad \Pi_M(G_1,G_2) \le G_1,$$

so that

$$\tau(\Pi_M(F_1,F_2),\Pi_M(G_1,G_2)) \le \tau(F_1,G_1);$$

in a similar manner,

$$\tau(\Pi_M(F_1,F_2),\Pi_M(G_1,G_2)) \le \tau(F_2,G_2),$$

whence

$$\tau(\Pi_M(F_1,F_2),\Pi_M(G_1,G_2))\Pi_M(\tau(F_1,G_1),\tau(F_2,G_2)),$$

viz., $\Pi_M \gg \tau$. □

Lemma 1.10.2. *For two t-norms T_1 and T_2 one has $T_1 \gg T_2$ if, and only if, $T_2^* \gg T_1^*$.*

Proof. Since for every t-norm $T, (T^*)^* = T$, one need only to prove that if $T_1 \gg T_2$ then $T_2^* \gg T_1^*$. Thus assume that for all F_1, F_2, G_1, and G_2 in Δ^+ and for all s, t, u, and v in $\mathbb{R}_+$, one has

$$\begin{aligned} &T_1[T_2(F_1(s),G_1(t)),T_2(F_2(u),G_2(v))] \\ &\qquad \ge T_2[T_1(F_1(s),F_2(u)),T_1(G_1(t),G_2(v))]. \end{aligned} \tag{1.10.2}$$

Then, by definition of t-conorm and by the preceding inequality,

$$\begin{aligned} &T_2^*[T_1^*(F_1(s),F_2(u)),T_1^*(G_1(t),G_2(v))] \\ &\quad = 1 - T_2[1 - T_1^*(F_1(s),F_2(u)), 1 - T_1^*(G_1(t),G_2(v))] \\ &\quad = 1 - T_2[T_1(1-F_1(s),1-F_2(u)),T_1(1-G_1(t),1-G_2(v))] \\ &\quad \ge 1 - T_1[T_2(1-F_1(s),1-G_1(t)),T_2(1-F_2(u),1-G_2(v))] \\ &\quad = T_1^*[T_2^*(F_1(s),G_1(t)),T_2^*(F_2(u),G_2(v))] \end{aligned}$$

i.e. the assertion. □

The following theorem, due to Tardiff (1975):

Theorem 1.10.1. *For two continuous t-norms T_1 and T_2 the following statements are equivalent:*

(a) $T_1 \gg T_2$;
(b) $\Pi_{T_1} \gg \Pi_{(T_2)}$;
(c) $\tau_{T_1} \gg \tau_{T_2}$;
(d) $\Pi_{T_1} \gg \tau_{T_2}$;
(e) $\tau_{T_2^*} \gg \tau_{T_1^*}$.

Proof. $(a) \implies (b)$ the inequality (1.10.2) is equivalent to (a) and this contains the special case $s = t = u = v$, which is (b).

$(b) \implies (c)$, again statement (b) is expressed by inequality (1.10.2) with $s = t = u = v$. □

1.11 Duality

The triangle function τ_M is constructed through the left-continuous t-norm M via

$$\tau_M(F, G)(x) := \sup\{M(F(u), G(v)) | Sum(u, v) = x\}. \tag{1.11.1}$$

If F and G are strict d.f.s, then the supremum on the right-hand side of (1.11.1) is attained precisely when $F(u) = G(v)$. Turning this observation around, we see that for any t in $\mathbb{I}$ there exist unique values u_t and v_t such that $F(u_t) = G(v_t) = t$ and $\tau_M(F, G)(u_t + v_t) = t$. Inverting, one has

$$[\tau_M(F, G)]^{-1}(t) = u_t + v_t = F^{-1}(t) + G^{-1}(t),$$

whence

$$[\tau_M(F, G)]^{-1} = F^{-1} + G^{-1}. \tag{1.11.2}$$

Display (1.11.2) remains valid for any F, G in Δ^+, (see Section 4.4 in (Schweizer & Sklar, 1983)) i.e., that

$$[\tau_M(F, G)]^\wedge = F^\wedge + G^\wedge, \tag{1.11.3}$$

from which we at once have

$$\tau_M(F, G) = [F^\wedge + G^\wedge]^\wedge \tag{1.11.4}$$

(see Sherwood & Taylor (1974, Prop. 4)). Display (1.11.4) shows that the operation τ_M in Δ^+ is equivalent to pointwise addition on the space of (left-continuous) quasi-inverses. Since the latter operation is simpler than the former, this is a useful result, applied in the sequel.

Next, since $F^\wedge$ and $G^\wedge$ are non-decreasing, we may write

$$F^\wedge(x) + G^\wedge(x) = \inf\{Sum(F^\wedge(u), G^\wedge(v)) | M(u,v) = x\}. \qquad (1.11.5)$$

The expressions on the right-hand sides of (1.11.1) and (1.11.5) are dual in the sense that each may be obtained from the other as follows: interchange M and Sum, interchange sup and inf, and replace functions by their quasi-inverses. Furthermore, (1.11.3) and (1.11.4) show that each expression is the quasi-inverse of the other. These observations, together with the definition of $\tau_{T,L}$ by

$$\tau_{T,L}(F,G)(x) = \sup\{T(F(u), G(v)) | L(u,v) = x\}, \qquad (1.11.6)$$

suggest that the forgoing relationships remain valid when M is replaced by any continuous t-norm T and Sum by any L. Frank and Schweizer (1979) showed that this is indeed generally the case — and in a much wider context. The results we need are summarized below.

Definition 1.11.1. For any F in Δ^+ let $F^\wedge$ be the left-continuous quasi-inverse of F. Then ∇ is the set $\{F^\wedge | F \in \Delta^+\}$.

Definition 1.11.2. For any T and any L, $\tau^\wedge_{T,L}$ is the function on $\nabla \times \nabla$ whose value for any $F^\wedge$, $G^\wedge$ in ∇^+ is the function $\tau^\wedge_{T,L}$ defined on $\mathbb{I}$ by

$$\tau^\wedge_{T,L}(F^\wedge, G^\wedge)(x) = \inf\{L(F^\wedge(u), G^\wedge(v)) | T(u,v) = x\}. \qquad (1.11.7)$$

If $L =$ Sum, then we write $\tau^\wedge_T$.

Theorem 1.11.1. *Suppose that L has ∞ as a null element and is continuous on all of $\mathbb{R}^+ \times \mathbb{R}^+$, and that T is continuous. Then for any F, G in Δ^+ we have*

$$\tau_{T,L}(F,G) = [\tau^\wedge_{T,L}(F^\wedge, G^\wedge)]^\wedge \qquad (1.11.8)$$

and

$$\tau^\wedge_{T,L}(F^\wedge, G^\wedge) = [\tau_{T,L}(F,G)]^\wedge. \qquad (1.11.9)$$

Thus $\tau^\wedge_{T,L}$ is a binary operation on ∇^+ that is non-decreasing in each place and has $\varepsilon^\wedge_0$ as identity.

If we let $T = M$ in (1.11.7), *then we obtain*

$$\tau^\wedge_{M,L}(F^\wedge, G^\wedge) = L(F^\wedge, G^\wedge), \qquad (1.11.10)$$

whence (1.11.8) *yields*

$$\tau_{M,L}(F,G) = [L(F^\wedge, G^\wedge)]^\wedge, \qquad (1.11.11)$$

and letting $L = Sum$ in (1.11.11) *brings us back to* (1.11.4).

Chapter 2

Probabilistic Normed Spaces

2.1 The First Definition

Šerstnev introduced the first definition of a probabilistic normed space in a series of papers (Šerstnev, 1962, 1963a, 1963b, 1964a); he was motivated by problems of best approximations in statistics. His definition runs along the same path followed in order to probabilize the notion of metric space and to introduce PM spaces.

Definition 2.1.1. A *probabilistic normed space* of Šerstnev (PN space) is a triple (V, ν, τ), where V is a (real or complex) linear space, ν is a mapping from V into Δ^+ and τ is a continuous triangle function and the following conditions are satisfied for all p and q in V

(N1) $\nu_p = \epsilon_0$ if, and only if, $p = \theta$ (θ is the null vector in V);
(N2) $\nu_{p+q} \geq \tau(\nu_p, \nu_q)$;

(Š) $\quad \forall \alpha \in \mathbb{R}\backslash\{0\}, \quad \forall x \in \overline{\mathbb{R}}_+ \quad \nu_{\alpha p}(x) = \nu_p\left(\frac{x}{|\alpha|}\right)$;

notice that condition (Š) implies

(N3) $\forall p \in V \quad \nu_{-p} = \nu_p$.

As an example of the kind of problem studied by Šerstnev, consider the following situation. Let (V, ν, τ) be a PN space, let A be a subset of V, and p a point not belonging to A. Is there a point $q \in A$ that is "closest" to p, in the sense that, for all $r \in A$, ν_{p-r} is not strictly greater than ν_{p-q}? If A is a finite dimensional subspace of V, then such a point q always exists (Theorem 4 in Šerstnev (1963b), and Theorem 5 in Šerstnev (1964b)). An extensive literature on PN spaces according to Šerstnev's definition was produced, mainly by the Russian school, in the 1960s and the 1970s; after that the topic seems to have lain dormant until 1993, when a new definition was proposed. This behavior contrasts sharply with what happened in the theory of PM spaces, which saw important developments. The reason for this is probably to be found in the fact that condition (Š) is very strong; it implies, for instance, that every one-dimensional subspace of a PN space is a simple PM space. Moreover, it has never been possible to formulate a reasonable definition of probabilistic inner product space.

2.2 1993: PN Spaces Redefined

A new, wider, definition of a PN space was introduced in Alsina *et al.* (1993). Their definition quickly became the standard one, and, to the best of the authors' knowledge, it has been adopted by all the researchers who, after them, have investigated the properties, the uses, or the applications of PN spaces. It is also the definition that will be adopted in this book. However, as will be seen, many cases are frequently encountered in which Šerstnev's definition is appropriate; this presents no problem since, as will be seen shortly, Šerstnev's definition is a particular case of the more general one.

The new definition is suggested by a result (Theorem 1 in Alsina *et al.* (1993)) that sheds light on the definition of a "classical" normed space.

Theorem 2.2.1. *Let V be a real linear space and let φ be a mapping from V into $\mathbb{R}_+$. Then the following are equivalent:*

(a) *for all $p \in V$ and all $\lambda \in \mathbb{R}$, $\varphi(\lambda p) = |\lambda|\varphi(p)$ (whence, in particular, $\varphi(\theta) = 0$, where θ is the null vector of the linear space V);*
(b) *the conditions*

$$\varphi(-p) = \varphi(p) \tag{2.2.1}$$

and, for all $p \in V$ and for all $\alpha \in [0, 1]$,

$$\varphi(p) = \varphi(\alpha p) + \varphi((1 - \alpha)p). \tag{2.2.2}$$

Thus the pair (V, φ) is a normed space if, and only if, (2.2.1), (2.2.2), *and the conditions*

$$\forall p \neq \theta \quad \varphi(p) \neq 0, \tag{2.2.3}$$

and

$$\forall p, q \in V \quad \varphi(p + q) \leq \varphi(p) + \varphi(q) \tag{2.2.4}$$

hold.

Proof. The implication $(a) \Rightarrow (b)$ is trivial.
$(b) \Rightarrow (a)$: For every $p \in V$, (2.2.2) yields

$$\varphi(p) = \varphi(0p) + \varphi(1p) = \varphi(0) + \varphi(p),$$

whence $\varphi(\theta) = 0$. Consequently, for all $p \in V$,

$$\varphi(0p) = \varphi(\theta) = 0 = 0\varphi(p).$$

The proof then proceeds by induction. Assume that there is a positive integer n such that $\varphi(np) = n\varphi(p)$ for all $p \in V$; then, on account of (2.2.2) one has

$$\begin{aligned}\varphi((n+1)p) &= \varphi\left(\frac{n}{n+1}(n+1)p\right) + \varphi\left(\frac{1}{n+1}(n+1)p\right) \\ &= \varphi(np) + \varphi(p) = (n+1)\varphi(p).\end{aligned}$$

Hence $\varphi(np) = n\varphi(p)$ for all $p \in V$ and for every positive integer $n \in \mathbb{Z}_+$. A standard argument (see p. 31 in Aczél (1966)) now gives $\varphi(rp) = r\varphi(p)$ for all $p \in V$ and for every positive rational $r \in \mathbb{Q}_+$. In fact, let r be a positive rational number, $r \in \mathbb{Q}_+$: then there are two positive integers m and n such that $r = \frac{m}{n}$. Thus $m = rn$ and

$$\varphi(mp) = m\varphi(p) \quad \text{and} \quad \varphi(mrp) = m\varphi(rp),$$

so that equality $\varphi(mp) = \varphi(nrp)$ yields

$$\varphi(rp) = \varphi\left(\frac{m}{n}p\right) = \frac{m}{n}\varphi(p).$$

Therefore one has $\varphi(rp) = r\varphi(p)$ for every rational $r \in \mathbb{Q}_+$ and for all $p \in V$ as asserted.

Next, if $0 \le \alpha < \beta$, then, for all $p \in V$,

$$\varphi(\beta p) = \varphi\left(\frac{\alpha}{\beta}\beta p\right) + \varphi\left(\frac{\beta-\alpha}{\beta}\beta p\right) = \varphi(\alpha p) + \varphi((\beta-\alpha)p) \ge \varphi(\alpha p).$$

Thus, for every $p \in V$ the mapping $\mathbb{R}_+ \ni \alpha \mapsto \varphi(\alpha p)$ is increasing. Now let α be a positive real number. There are two sequences of positive rational numbers $\{r_n\}$ and $\{s_n\}$, such that for every $n \in \mathbf{N}$ one has $r_n \le \alpha \le s_n$, and

$$\lim_{n\to+\infty} r_n = \lim_{n\to+\infty} s_n = \alpha.$$

Since $\alpha \mapsto \varphi(\alpha p)$ is increasing for every $p \in V$, one has

$$r_n\varphi(p) = \varphi(r_n p) \le \varphi(\alpha p) \le \varphi(s_n p) = s_n\varphi(p).$$

Consequently, letting $n \to +\infty$, one has $\varphi(\alpha p) = \alpha\varphi(p)$ for all $p \in V$ and for every real number $\alpha \geq 0$. Finally, an application of (2.2.1) yields $\varphi(\alpha p) = \alpha\varphi(p)$ for all $p \in V$ and for every real number α. □

Applying Theorem 2.2.1 yields the following result, which clarifies the role of the Šerstnev condition (Š) of Definition 2.1.1.

Theorem 2.2.2. *For a pair (V, ν) that satisfies conditions $(N1)$ and $(N2)$ the following statements are equivalent:*

(a) *(V, ν) satisfies also condition (Š);*
(b) *for all $p \in V$ and for all $\alpha \in [0, 1]$*

$$\nu_p = \tau_M(\nu_{\alpha p}, \nu_{(1-\alpha)p}). \tag{2.2.5}$$

Proof. For every $F \in \Delta^+$ let $F^{[-1]}$ denote the left continuous quasi-inverse[1] of F. Having recourse to duality (see Section 1.11), for all F, G, and H in Δ^+ $H = \tau_M(F, G)$ if, and only if,

$$H^{[-1]} = F^{[-1]} + G^{[-1]}.$$

Therefore (2.2.5) holds if, and only if, for all $p \in V$ and all $\alpha \in [0, 1]$,

$$\nu_p^{[-1]} = \nu_{\alpha p}^{[-1]} + \nu_{(1-\alpha)p}^{[-1]}. \tag{2.2.6}$$

$(a) \Rightarrow (b)$ For all $p \in V$, for every $\alpha \in [0, 1]$, and for every $t \in \mathbb{R}_+$

$$\nu_{\alpha p} = \nu_p\left(\frac{t}{\alpha}\right) \quad \text{and} \quad \nu_{(1-\alpha)p}(t) = \nu_p\left(\frac{t}{1-\alpha}\right).$$

It follows from the definition of quasi-inverse that

$$\nu_{\alpha p}^{[-1]} = \alpha\nu_p^{[-1]} \quad \text{and} \quad \nu_{(1-\alpha)p}^{[-1]} = (1-\alpha)\nu_p^{[-1]}, \tag{2.2.7}$$

so that (2.2.6) holds for every $\alpha \in [0, 1]$. Since (2.2.6) holds simultaneously for $\alpha = 0$ and for $\alpha = 1$, by virtue of (N1), (2.2.6) holds for all $\alpha \in [0, 1]$, whence (2.2.5) holds.

$(a) \Rightarrow (b)$ It follows from (2.2.6) that the function $f_t : V \to \mathbb{R}_+$ defined for a given $t \in [0, 1]$ by $f_t(p) := \nu_p^{[-1]}(t)$ satisfies Eqs (2.2.1) and (2.2.2). Therefore, for every $\alpha \in \mathbb{R}_+$ and every $t \in [0, 1]$, the first of the (2.2.7) gives

$$\nu_{\alpha p}^{[-1]}(t) = f_t(\alpha p) = |\alpha| f_t(p) = |\alpha|\nu_p^{[-1]}(t),$$

whence $\nu_{\alpha p}^{[-1]} = |\alpha|\nu_p^{[-1]}$, which is equivalent to (Š). □

[1] Notice that throughout this book $[-1]$ and $\wedge$ are synonymous with quasi-inverse.

Therefore if (S, ν) satisfies (N1), (N2), and (N3), then it is a Šerstnev space if, and only if, (2.2.5) holds for all $p \in V$ and for all $\alpha\,[0,1]$; in this case one has, for all $p \in V$ and for all $\alpha \in [0,1]$,

$$\tau(\nu_{\alpha p}, \nu_{(1-\alpha)p}) \leq \nu_p = \tau_M(\nu_{\alpha p}, \nu_{(1-\alpha)p}).$$

The definition of PN space then runs as follows:

Definition 2.2.1. A *probabilistic normed space,* which will henceforth be called briefly a PN space, is a quadruple (V, ν, τ, τ^*), where V is a linear space, τ and τ^* are continuous triangle functions, and the mapping $\nu : V \to \Delta^+$ satisfies, for all p and q in V, the conditions

(N1) $\nu_p = \epsilon_0$ if, and only if, $p = \theta$ (θ is the null vector in V);
(N2) $\forall p \in V \ \nu_{-p} = \nu_p$;
(N3) $\nu_{p+q} \geq \tau(\nu_p, \nu_q)$;
(N4) $\forall \alpha \in [0,1] \ \nu_p \leq \tau^*(\nu_{\alpha p}, \nu_{(1-\alpha)p})$.

The function ν is called the *probabilistic norm.* If (V, ν, τ, τ^*) satisfies the condition, weaker than (N1),

$$\nu_\theta = \epsilon_0,$$

then it is called a *probabilistic pseudo-normed space* (PPN space). If $\tau = \tau_T$ and $\tau^* = \tau_{T^*}$ for some continuous t-norm T and its t-conorm T^* then $(V, \nu, \tau_T, \tau_{T^*})$ is denoted by (V, ν, T) and is a *Menger PN space.*

Since $\tau_M = \tau_{M^*}$ (see Corollary 7.5.8 in Schweizer and Sklar (1983)) the notions of Šerstnev and Menger PN spaces coincide when $\tau = \tau_M$. In general, since $\tau_M \leq \tau_{T^*}$ for every t-norm T, a Šerstnev PN space (V, ν, τ_T) is also a Menger PN space. The converse need not hold as shown by the following example (Alsina, Schweizer, and Sklar, 1993).

Example 2.2.1. Define $\nu : \mathbb{R} \to \Delta^+$ by $\nu_\theta = \epsilon_0$ and, if $p \neq \theta$ by

$$\nu_p(t) = \begin{cases} 0, & t \leq 0, \\ exp(-\sqrt{|p|}), & t \in]0, +\infty[, \\ 1, & t = +\infty. \end{cases}$$

Then $(\mathbb{R}, \nu)$ is a Menger space under Π. In fact, for every $t \in]0, +\infty[$, one has

$$\tau_\Pi(\nu_p, \nu_q)(t) = \sup_{s\in[0,t]} \{\nu_p(s), \nu_q(t-s)\} = exp\{-\sqrt{|p|}\}exp\{-\sqrt{|q|}\}.$$

It follows immediately from the triangle inequality $|p+q| \leq |p| + |q|$ that

$$\tau_\Pi(\nu_p, \nu_q)(t) \leq \nu_{p+q}(t),$$

namely (N3). Since $\Pi^*(x, y) = x + y - xy$, one has, for every $t \in]0, +\infty[$ and for every $\alpha \in [0, 1]$,

$$\begin{aligned}\tau_{\Pi^*}(\nu_{\alpha p}, \nu_{(1-\alpha)p})(t) &= \inf_{s\in[0,\,t]} \{\nu_{\alpha p}(s) + \nu_{(1-\alpha)p}(t-s) - \nu_{\alpha p}(s)\nu_{(1-\alpha)p}(t-s)\} \\ &= exp\{-\sqrt{\alpha|p|}\} + exp\{-\sqrt{(1-\alpha)|p|}\} \\ &\quad - exp\{-|p|\sqrt{\alpha(1-\alpha)}\}.\end{aligned}$$

However, if $t \in]0, +\infty[, \alpha \in]0, 1[$ and if $p \neq 0$ one has

$$\begin{aligned}\nu_p(t) &= exp\{-\sqrt{|p|}\} < exp\{-\sqrt{\alpha|p|}\} \wedge exp\{-\sqrt{(1-\alpha)|p|}\} \\ &= \tau_M(\nu_{\alpha p}, \nu_{(1-\alpha)p})\end{aligned}$$

so that (S, ν, τ_Π) is not a Šerstnev PN space.

Let $(V, \|\cdot\|)$ be a normed space; notice that any quadruple (V, ν, τ, τ^*) is a PN space, if $\nu : V \to \Delta^+$ by

$$\nu_p := \epsilon_{\|p\|}$$

and if the triangle functions τ and τ^* are such that, for all a and b in $\mathbb{R}_+$,

$$\tau(\epsilon_a, \epsilon_b) \leq \epsilon_{a+b} \leq \tau^*(\epsilon_a, \epsilon_b). \tag{2.2.8}$$

As a consequence, the family of real normed spaces is strictly included in the set of PN spaces. Notice in particular that, for every continuous t-norm T, the stricter relation

$$\tau_T(\epsilon_a, \epsilon_b) = \epsilon_{a+b} = \tau_{T^*}(\epsilon_a, \epsilon_b) \tag{2.2.9}$$

holds.

In the sequel many examples of PN spaces will be encountered; however, the following is the simplest possible example of a PN space.

Example 2.2.2. Let $(\mathbb{R}, |\cdot|)$ be the set of all real numbers endowed with the usual norm. Since for every continuous t-norm T(2.2.9) holds, $(\mathbb{R}, |\cdot|)$ is a Menger PN space under any continuous t-norm T.

Given a real normed space $(V, \|\cdot\|)$, there is always a PN space associated to it. Define a map $\nu^c : V \to \Delta^+$ through

$$\nu_p^c(t) = \begin{cases} \dfrac{t}{t + \|p\|}, & \text{if } t \in]0, +\infty[, \\ 1, & \text{if } t = +\infty. \end{cases} \tag{2.2.10}$$

Then one has the following result. By Π_Π and Π_{Π^*} we denote, respectively, the binary operations on Δ^+ defined, for all F and G in Δ^+ and every $t \geq 0$, via

$$\Pi_\Pi(F, G)(t) := F(t)G(t), \quad \text{and} \quad \Pi_{\Pi^*}(F, G)(t) := F(t) + G(t) - F(t)G(t).$$

Theorem 2.2.3. *For every real normed space* $(V, \|\cdot\|)$, *the quadruple* $(V, \nu^c, \Pi_\Pi, \Pi_{\Pi^*})$ *is a PN space.*

Proof. Properties (N1) and (N2) are obvious. (N3) follows from the inequality

$$(t + \|p\|)(t + \|q\|) \geq t(t + \|p + q\|),$$

which holds for all p and q in V. As for (N4), the inequality to be proved is equivalent, after a few easy calculations, to the other one

$$\alpha(1 - \alpha)\|p\|^2 \leq \|p\|^2 + \|p\|t \quad \text{and} \quad \alpha(1 - \alpha)\|p\| \leq \|p\| + t,$$

which is obviously true. □

The PN space just introduced $(V, \nu^c, \Pi_\Pi, \Pi_{\Pi^*})$ will be called the *canonical PN space associated with the normed space* $(V, \|\cdot\|)$.

2.3 Special Classes of PN Spaces

In the next few sections we present examples of PN spaces: for this we refer to Lafuerza-Guillen, Rodriguez Lallena, and Sempi (1997). In this section, F will denote a d.f. different from either ε_0 or ε_∞.

2.3.1 *Equilateral spaces*

Define $\nu : V \to \Delta^+$ by $\nu_\theta := \varepsilon_0$, while, if $p \neq \theta, \nu_p := F$. Then (V, ν) is a PPN space under the triangle function Π_M; it will be denoted by (V, F, Π_M).

2.3.2 *Simple PN spaces*

Definition 2.3.1. Let $(V, \|\cdot\|)$ be a normed space and define $\nu : V \to \Delta^+$ by $\nu_\theta := \varepsilon_0$, and, if $p \neq \theta$, by

$$\nu_p(t) := F\left(\frac{t}{\|p\|}\right) \quad (t > 0),$$

where F is a d.f. different from ε_0 or ε_∞. The pair (V, ν) is called the *simple space* generated by $(V, \|\cdot\|)$ and F.

Theorem 2.3.1. *The simple space (V, ν) generated by $(V, \|\cdot\|)$ and F is a Menger PN space under M, denoted by $(V, \|\cdot\|, F, M)$ and a Šerstnev space.*

Proof. Let $\nu_p := \varepsilon_0$ and assume, if possible, that $p \neq \theta$; therefore, for every $t > 0$, one has $F(\frac{t}{\|p\|}) = 1$. Since $\|p\| > 0$, this would imply $F = \varepsilon_0$, contrary to the assumption. This proves (N1). Property (N2) is obvious. In order to prove (N3) we shall have recourse to the duality introduced in Frank & Schweizer (1979) (see also Section 7.7 in Schweizer and Sklar (1983)). Let $G^{[-1]}$ denote the quasi-inverse of the distance d.f. G. Since $\nu_p^{[-1]} = \|p\| G^{[-1]}$, one has, for all p and q in V

$$\begin{aligned}[\tau_M(\nu_p, \nu_q)]^{[-1]} &= \nu_p^{[-1]} + \nu_{q^{[-1]}} = \|p\| G^{[-1]} + \|q\| G^{[-1]} \\ &= (\|p\| + \|q\|) G^{[-1]} \geq \|p + q\| G^{[-1]} = \nu_p^{[-1]}\end{aligned}$$

so that $\nu_{p+q} \geq \tau_M(\nu_p, \nu_q)$, *viz.* property (N3) holds.

In order to prove (N4) we shall use the equality $\tau_{M^*} = \tau_M$ (see Corollary 7.5.8 in Schweizer and Sklar (1983)). Thus the argument just used yields, for every $\alpha \in [0, 1]$,

$$\begin{aligned}[\tau_{M^*}(\nu_{\alpha p}, \nu_{(1-\alpha)p})]^{[-1]} &= [\tau_M(\nu_{\alpha p}, \nu_{(1-\alpha)p}]^{[-1]} \\ &= \alpha\|p\| G^{[-1]} + (1-\alpha)\|p\| G^{[-1]} = \|p\| G^{[-1]} = \nu_{p+q}^{[-1]},\end{aligned}$$

Hence the assertion. By virtue of Theorem 2.2.2, a simple space is a Šerstnev space under τ_M. □

2.4 α-simple Spaces

Definition 2.4.1. If $\alpha \geq 0$ define $\nu : V \to \Delta^+$ by $\nu_\theta := \varepsilon_0$, and, if $p \neq \theta$, by

$$\nu_p(t) := F\left(\frac{t}{\|p\|^\alpha}\right) \quad (t > 0),$$

where F is a d.f. different from ε_0 and ε_∞. The pair (V, ν) is called the α-*simple space* generated by $(V, \|\cdot\|)$ and F.

It is immediately seen that the *α-simple space* generated by $(V, \|\cdot\|)$ and F is a PSN, which will be denoted by $(V, \|\cdot\|, F; \alpha)$. The PSM space associated with $(V, \|\cdot\|, F; \alpha)$ is the *α-simple PSM space* $(V, d, F; \alpha)$, where d is the metric of the norm $\|\cdot\|$ i.e. $d(p,q) := \|p - q\|$. For $\alpha = 0$ and $\alpha = 1$ one obtains the equilateral and the simple PN spaces respectively.

In the case $\alpha \in]0, 1[$ it is instructive to compare the different behavior of the PSM $(V, d^\alpha, F; \alpha)$ and of the PSN $(V, \|\cdot\|, F; \alpha)$. In fact, d^α is a metric, so that $(V, d^\alpha, F; \alpha)$ is a Menger PN space under M (see Theorem 8.4.2 in Schweizer and Sklar (1983)). But $\|\cdot\|^\alpha$ is not a norm if $\alpha \in]0, 1[$, so that $(V, \|\cdot\|, F; \alpha)$ need not be a Menger PN space under M, as the following example shows.

Example 2.4.1. Consider the *α-simple space* $(V, \|\cdot\|, U; \alpha)$, where $\alpha \in]0, 1[$ and U is the d.f. of the distribution on $(0, 1)$. Then axiom (N4) does not hold for $\lambda = \frac{1}{2}$. In fact $\nu_p(\|p\|^\alpha) = 1$, while, for every $F \in \Delta^+$ and for every $t \geq 0$, one has

$$\tau_{M^*}(F, F)(t) = \tau_M(F, F)(t) = F\left(\frac{1}{2}\right).$$

Therefore,

$$\begin{aligned}\tau_{M^*}(\nu_{p/2}, \nu_{p/2})(\|p\|^\alpha) &= U\left(\frac{\|p\|^\alpha}{2\|p/2\|^\alpha}\right) = \frac{2^\alpha}{\|p\|^\alpha}\frac{\|p\|^\alpha}{2} \\ &= \frac{1}{2^{1-\alpha}} < 1 = \nu_p(\|p\|^\alpha).\end{aligned}$$

It is not hard to deduce from Theorem 8.6.2 in Schweizer & Sklar (1983), which holds with the same proof, that an *α-simple PSN space* with $\alpha > 1$ need not be a Menger space under M. We recall that a large class of *α-simple PSM spaces* can be endowed with the structure of a Menger space (see Schweizer and Sklar, 1963) or Theorem 8.6.5 in Schweizer and Sklar (1983). The next theorems show that analogous results hold for PSN. However, their proofs are not trivial extensions of the respective ones for *α-simple PSM spaces*; moreover, these results will also have to be proved for the case $\alpha \in]0, 1[$.

The main results for *α-simple PN spaces* are given in Theorems 2.4.1 and 2.4.2.

By straightforward calculation one gets:

Lemma 2.4.1. *Let $(V, \|\cdot\|)$ be a normed space, $F \in \mathcal{D}^+$ a strictly increasing continuous d.f., T a strict t-norm with additive generator f, and $\alpha > 0$*

with $\alpha \neq 1$. Then the following statements are equivalent:

(a) $(V, \|\cdot\|, F; \alpha)$ *is a Menger space under* T;
(b) *the following inequalities hold for all* u *and* v *in* $]0, \infty[$, *for every pair of points* p *and* q *in* V, *with* $p \neq \theta, q \neq \theta$ *and* $p + q \neq \theta$,

$$(f \circ F)\left(\frac{u+v}{\|p+q\|}\right) \leq (f \circ F)\left(\frac{u}{\|p\|^\alpha}\right) + (f \circ F)\left(\frac{v}{\|q\|^\alpha}\right) \tag{2.4.1}$$

and

$$(f \circ F^*)\left(\frac{u+v}{\|p\|^\alpha}\right) \leq (f \circ F^*)\left(\frac{u}{\lambda^\alpha \|p\|^\alpha}\right) + (f \circ F^*)\left(\frac{v}{(1-\lambda)^\alpha \|q\|^\alpha}\right) \tag{2.4.2}$$

where for $t > 0$, $F^*(t) := 1 - F(t)$.

Set $h := f \circ F$ and $h^* := f \circ F^*$ to obtain

(a) both h and h^* map $[0, +\infty]$ into $[0, +\infty]$, $h(0) = h^*(+\infty), h(+\infty) = h^*(0) = 0$;
(b) both h and h^* are continuous;
(c) h is strictly decreasing and h^* is strictly increasing.

Therefore their inverse h^{-1} and $(h^*)^{-1}$ have the same properties as h and h^* respectively.

Let p and q be in V, with $p \neq \theta, q \neq \theta$ and $p + q \neq \theta$ and let $\lambda \in [0, 1]$. For $u, v > 0$ define

$$s := h\left(\frac{u}{\|p\|^\alpha}\right) \quad \textit{and} \quad t := h\left(\frac{v}{\|q\|^\alpha}\right)$$

thus $h^{-1}(s) = \frac{u}{\|p\|^\alpha}$ and $h^{-1}(t) = \frac{v}{\|q\|^\alpha}$. Now the display (2.4.1) is

$$h\left(\frac{\|p\|^\alpha h^{-1}(s) + \|q\|^\alpha h^{-1}(t)}{\|p+q\|^\alpha}\right) \leq s + t$$

equivalent to

$$\|p+q\|^\alpha h^{-1}(s+t) \leq \|p\|^\alpha h^{-1}(s) + \|q\|^\alpha h^{-1}(t).$$

In a similar way, if we define

$$s = h^*\left(\frac{u}{\lambda^\alpha \|p\|^\alpha}\right) \quad \textit{and} \quad t = h^*\left(\frac{v}{(1-\lambda)^\alpha \|q\|^\alpha}\right)$$

one proves that (2.4.2) is equivalent to

$$\lambda^\alpha (h^*)^{-1}(s) + (1-\lambda)^\alpha (h^*)^{-1}(t) \leq (h^*)^{-1}(s+t).$$

Thus the Lemma 2.4.1 takes the following form.

Lemma 2.4.2. *Let $(V, \|\cdot\|)$ be a normed space, $F \in \mathcal{D}^+$ a strictly increasing continuous d.f., T a strict t-norm with additive generator f, and $\alpha > 0$ with $\alpha \neq 1$. Then the following statements are equivalent:*

(a) *$(V, \|\cdot\|, F; \alpha)$ is a Menger space under T;*
(b) *The following inequalities hold for all s and t in $]0, +\infty[$, for every $\lambda \in [0, 1]$ and for every pair of points p and q in V, with $p \neq \theta, q \neq \theta$ and $p + q \neq \theta$,*

$$\|p+q\|^\alpha (f \circ F)^{-1}(s+t) \leq \|p\|^\alpha (f \circ F)^{-1}(s) + \|q\|^\alpha (f \circ F)^{-1}(t) \tag{2.4.3}$$

and

$$(f \circ F)^{-1}(s+t) \geq \lambda^\alpha (f \circ F)^{-1}(s) + (1-\lambda)^\alpha (f \circ F)^{-1}(t) \tag{2.4.4}$$

where, for $t > 0, F^*(t) := 1 - F(t)$.

When $\alpha < 1$, inequality (2.4.3) is trivially satisfied, because $\|p + q\|^\alpha \neq \|p\|^\alpha + \|q\|^\alpha$ and $(f \circ F)^{-1}$ is strictly decreasing. On the other hand, inequality (2.4.4) is trivial when $\alpha < 1$, since $(f \circ F)^{-1}$ is strictly increasing and, for every $\lambda \in]0, +\infty[, \lambda^\alpha + (1 - \lambda)^\alpha < 1$. As a consequence one can rephrase Lemma 2.4.2 in the following form.

Lemma 2.4.3. *Let $(V, \|\cdot\|)$ be a normed space, $F \in \mathcal{D}^+$ a strictly increasing continuous d.f., T a strict t-norm with additive generator f and $F^*(t) := 1 - F(t)$.*

(a) *If $\alpha \in]0, 1[$, then $(V, \|\cdot\|, F; \alpha)$ is a Menger space under T if, and only if, for every $\lambda \in]0, 1[$ and for all $s, t \in]0, +\infty[$, inequality (2.4.4) holds.*
(b) *If $\alpha \in]1, +\infty[$, then $(V, \|\cdot\|, F; \alpha)$ is a Menger space under T if, and only if, for all $s, t \in]0, +\infty[$ and for all $p, q \in V$, with $p \neq \theta, q \neq \theta$ and $p + q \neq \theta$, inequality (2.4.3) holds.*

Lemma 2.4.4. *If, beside the conditions of the previous lemma, one has, for every $\alpha > 1$,*

$$(f \circ F)(t) = t^{1/(1-\alpha)}, (t \in [0, +\infty]) \tag{2.4.5}$$

or, for $\alpha \in]0, 1[$

$$(f \circ F^*)(t) = t^{1/(1-\alpha)}, (t \in [0, +\infty]) \tag{2.4.6}$$

then $(V, \|\cdot\|, F; \alpha)$ is a Menger space under T.

Proof. In view of (2.4.5), if $\alpha > 1$ and $s, t \in]0, +\infty[$ then inequality (2.4.3) is implied, for all $p, q \in V$, with $p \neq \theta, q \neq \theta$ and $p + q \neq \theta$, the following inequality

$$(s+t)^{1-\alpha} \leq x^\alpha s^{1-\alpha} + (1-x)^\alpha t^{1-\alpha}$$

which holds for every $x \in]0,1[$ and which can be proved in a straightforward manner. Similarly, in view of (2.4.6) if $\alpha \in]0,1[$ one can prove that, for every $x \in]0,1[$ and for all $s, t \in]0, +\infty[$, the inequality

$$(s+t)^{1-\alpha} \geq x^\alpha s^{1-\alpha} + (1-x)^\alpha t^{1-\alpha}$$

holds; from the inequality (2.4.4) follows by setting $x = \lambda$. □

The main results can be stated now.

Theorem 2.4.1. *Let $(V, \|\cdot\|)$ be a normed space and let $\alpha > 1$,*

(a) *If $F \in \mathcal{D}^+$ is continuous and strictly increasing, then $(V, \|\cdot\|, F; \alpha)$ is a Menger space under the strict t-norm defined, for all x and y in $[0, +\infty]$, by*

$$T_F(x,y) := F\left(\{[F^{-1}(x)]^{1/(1-\alpha)} + [F^{-1}(y)]^{1/(1-\alpha)}\}^{1-\alpha}\right); \qquad (2.4.7)$$

(b) *If T is a strict t-norm with additive generator f, then the function $F : [0, +\infty] \to [0,1]$ defined by*

$$F(t) := f^{-1}(x^{1/(1-\alpha)})$$

is a continuous, strictly increasing d.f. of $\mathcal{D}^+$ and $(V, \|\cdot\|, F; \alpha)$ is a Menger space under T.

Proof. (a) Let F satisfy the assumptions; then define, for $x \in [0,1]$,

$$f(t) := [F^{-1}(t)]^{1/(1-\alpha)}.$$

The function thus defined is the additive generator of a strict t-norm in fact, f is strictly increasing, $f^{(\alpha-1)}(1) = 0$, since $F^{-1}(1) = +\infty$ and $f^{(\alpha-1)}(0) = +\infty$ since $F^{-1}(+\infty) = 1$. The strict t-norm in question is given by (2.4.7). The final assertion follows from Lemma 2.4.4.

(b) Given the strict t-norm T having f as additive generator, define

$$F(t) := f^{-1}(x^{1/(1-\alpha)}) \quad (t \in]0, +\infty[).$$

The function F thus defined is a continuous strictly increasing d.f. in $\mathcal{D}^+$; again the assertion follows from Lemma 2.4.4. □

Theorem 2.4.2. *Let $(V, \|\cdot\|)$ be a normed space and let $\alpha > 1$,*

(a) *If $F \in \mathcal{D}^+$ is continuous and strictly increasing, then $(V, \|\cdot\|, F; \alpha)$ is a Menger space under the strict t-norm defined, for all x and y in $[0, +\infty]$, by*

$$T_{F^*}(x, y) := F^*(\{[(F^*)^{-1}(x)]^{1/(1-\alpha)} + [(F^*)^{-1}(y)]^{1/(1-\alpha)}\}^{1-\alpha}); \tag{2.4.8}$$

where $F^*(t) = 1 - F(t)$;

(b) *If T is a strict t-norm with additive generator f, then the function $F : [0, +\infty] \to [0, 1]$ defined by*

$$F(t) := 1 - f^{-1}(x^{1/(1-\alpha)})$$

is a continuous, strictly increasing d.f. of $\mathcal{D}^+$ and $(V, \|\cdot\|, F; \alpha)$ is a Menger space under T.

Proof. (a) Let F satisfy the assumptions; then define, for $x \in [0, 1]$,

$$f(t) := [(F^*)^{-1}(t)]^{1/(1-\alpha)}$$

The function thus defined is the additive generator of a strict t-norm T_{G^*}; the proof of this fact is very similar to that of the previous theorem. Moreover, $(f \circ F^*)(t) = t^{1/(1-\alpha)}$ so that the assertion is a consequence of Lemma 2.4.4.

(b) Given the strict t-norm T having f as additive generator, define

$$F(t) := 1 - f^{-1}(x^{1/(1-\alpha)}) \quad (t \in]0, +\infty[).$$

Then F is a continuous strictly increasing d.f. in $\mathcal{D}^+$, $F^*(t) = f^{-1}(t^{1/(1-\alpha)})$ and, as a consequence $(f \circ F^*)(t) = t^{1/(1-\alpha)}$ so that the assertion follows from Lemma 2.4.4. □

The following two results are the analogue for PN spaces of Theorem 3 in Schweizer and Sklar (1983) valid for PM spaces. They also establish the relevance of the t-norms T_F and T_{F^*} of Theorems 2.4.1 and 2.4.2.

Theorem 2.4.3. *For every $\alpha > 1$ there exist normed spaces $(V, \|\cdot\|)$ with the following properties:*

(a) *If $F \in \mathcal{D}^+$ is continuous and strictly increasing, then T_F is the strongest continuous t-norm under which $(V, \|\cdot\|, F; \alpha)$ is a Menger space, in the sense that, if T is any other continuous t-norm that makes it a PN space, then $T \leq T_F$;*

(b) *If T is a strict t-norm with additive generator f, then T is the strongest t-norm under which $(V, \|\cdot\|, F; \alpha)$, with*

$$F(t) := f^{-1}(t^{1/(1-\alpha)})$$

is a Menger PN space.

Proof. (a) Take as the normed space $(V, \|\cdot\|)$ the set of the real numbers endowed with the Euclidean norm, $(\mathbb{R}, |\cdot|)$. Assume that $(\mathbb{R}, |\cdot|, F; \alpha)$, is a Menger PN space under the continuous t-norm T; then it has to be proved that $T(s,t) \leq T_F(s,t)$ at every point $(s,t) \in [0,1]^2$. In fact, one only needs to prove this inequality in the interior of $[0,1]^2$. Now let $(s,t) \in [0,1]^2$ be fixed and set

$$p := [g^{-1}(s)]^{1/(1-\alpha)} \quad \text{and} \quad q := [g^{-1}(t)]^{1/(1-\alpha)},$$

Both p and q belong to $]0, +\infty[$ and

$$\nu_p(p) = F\left(\frac{p}{|p|^\alpha}\right) = F(p^{1-\alpha}) = F(F^{-1}(s)) = s;$$

similarly, one has $\nu_q(q) = t$ so that

$$\begin{aligned} T(s,t) &= T(\nu_p(p), \nu_q(q)) \leq \sup_{u+v=p+q} \{T(\nu_p(u), \nu_q(q))\} \\ &= \tau_T(\nu_p, \nu_q)(p+q) \leq \nu_{p+q}(p+q) = F\left(\frac{p+q}{|p+q|^\alpha}\right). \end{aligned}$$

On the other hand

$$\begin{aligned} T_F(s,t) &= T_F(\nu_p(p), \nu_q(q)) \\ &= F\{([F^{-1}(\nu_p(p))]^{1/(1-\alpha)} + [F^{-1}(\nu_q(q))]^{1/(1-\alpha)})^{1-\alpha}\} \\ &= F((p+q)^{1-\alpha})) = F\left(\frac{p+q}{|p+q|^\alpha}\right). \end{aligned}$$

This completes the proof of (a). The same example proves (b). □

Theorem 2.4.4. *For every $\alpha \in [0,1]$ there exists normed space $(V, \|\cdot\|)$ with the following properties:*

(a) *If $F \in \mathcal{D}^+$ is continuous and strictly increasing, then T_{F^*} is the strongest continuous t-norm under which $(V, \|\cdot\|, F; \alpha)$ is a Menger PN space;*

(b) *If T is a strict t-norm with additive generator f, then T is the strongest t-norm under which $(V, \|\cdot\|, F; \alpha)$, with*

$$F(t) := f^{-1}(t^{1/(1-\alpha)})$$

is a Menger PN space.

Proof. (a) As in the previous proof, assume that $(\mathbb{R}, |\cdot|, F; \alpha)$ is a Menger PN space under the t-norm T. For every point $(s,t) \in [0,1]^2$, set

$$p := [(F^*)^{-1}(s)]^{1/(1-\alpha)} \quad q := [(F^*)^{-1}(t)]^{1/(1-\alpha)} \quad \lambda = \frac{p}{p+q};$$

then p and q are in $]0, +\infty[$ and $\lambda \in]0,1[$.

Then

$$\begin{aligned}
\nu_{\lambda(p+q)}(\lambda(p+q)) &= F\left(\frac{\lambda(p+q)}{(\lambda(p+q))^{\lambda}}\right) = F\left((\lambda(p+q))^{1-\alpha}\right) \\
&= F(p^{1-\alpha}) = F((F^*)^{-1}(s)) \\
&= 1 - F^*((F^*)^{-1}(s)) = 1 - s.
\end{aligned}$$

Similarly, $\nu_{1-\lambda(p+q)}((1-\lambda)(p+q)) = 1 - t$. Therefore

$$\begin{aligned}
T(s,t) &= 1 - T^*(1-s, 1-t) \\
&= 1 - T^*(\nu_{\lambda(p+q)}(\lambda(p+q), \nu_{(1-\lambda)(p+q)}((1-\lambda)(p+q)) \\
&\leq 1 - \inf_{u+v=p+q} \{T^*(\nu_{\lambda(p+q)}(u, \nu_{(1-\lambda)(p+q)})(v)\} \\
&= 1 - \tau_{T^*}(\nu_{\lambda(p+q)}, \nu_{(1-\lambda)(p+q)})(p+q) \leq 1 - \nu_{p+q}(p+q) \\
&= 1 - F\left(\frac{p+q}{(p+q)^{\alpha}}\right) = F^*((p+q)^{1-\alpha}).
\end{aligned}$$

On the other hand

$$T_{F^*}(s,t) = T_{F^*}(F^*(p^{1-\alpha}), F^*(q^{1-\alpha})) = F^*((p+q)^{1-\alpha}).$$

Part (b) is proved in the same way. □

Finally we show that an α-simple PN space is not, in general, a Šerstnev space.

Theorem 2.4.5. *Let $\alpha > 0$ and $\alpha \neq 1$. The following statements are equivalent for an α-simple space $(V, \|\cdot\|, F; \alpha)$:*

(a) *$(V, \|\cdot\|, F; \alpha)$ is a Šerstnev PSN space under some triangle function τ;*
(b) *F is constant on $]0, +\infty[$.*

Proof. Since the implication $(b) \Rightarrow (a)$ is immediate, one need only deal with the order implication $(a) \Rightarrow (b)$.

Axiom $(\hat{S})$ yields, for all $p \neq \theta, \lambda \neq 0$ and $t > 0$,

$$F\left(\frac{t}{\|\lambda p\|^{\alpha}}\right) = F\left(\frac{t}{|\lambda|\|p\|^{\alpha}}\right),$$

which is equivalent to saying $F(x) = F(|\lambda|^{1-\alpha}x)$ for all $x > 0$ and $\lambda \neq 0$; this implies that F is constant on $]0, +\infty[$. □

2.5 EN Spaces

EN spaces, shortly to be defined, provide an important class of PN spaces. Their importance derives from the role they play in the study of convergence of random variables.

Definition 2.5.1. Let $(\Omega, \mathcal{A}, P)$ be a probability space, $(V, \|\cdot\|)$ be a normed space and S a linear space of V-valued random variables (possibly the entire space). For every $p \in S$ and for every $t \in \overline{\mathbb{R}}_+$, define a mapping $\nu : S \to \Delta^+$ via

$$\nu_p(t) := P(\{\omega \in \Omega : \|p(\omega)\| < t\}) \tag{2.5.1}$$

where P is a *probability measure* on Ω. The couple (S, ν) is called an E-normed space (EN space).

Such spaces have been introduced by Sherwood (1969, 1979). The results that follow have been established in Lafuerza-Guillén, Rodríguez Lallena and Sempi (1977) and Lafuerza-Guillén and Sempi (2003). For the sake of completeness the complete proof will be reproduced adapting it from Sherwood (1969, 1979) although the new part, which is necessary in order to extend the proof to PN spaces, is that regarding axiom (N4).

Definition 2.5.2. An EN space (S, ν) is said to be canonical if it is a PN space under the two triangle functions τ_W and τ_M.

Theorem 2.5.1. *An EN space (S, ν) is a PPN space under the triangle functions τ_W and τ_M.*

Proof. Only the proof of the properties (N3) and (N4) is needed. For all p, q, and r in S and for every $t > 0$, let u and v in $[0, +\infty]$ be such that

$u + v = t$. Define the sets A, B, and C by

$$A = \{\omega \in \Omega : \|p(\omega)\| < u\} \quad B = \{\omega \in \Omega : \|q(\omega)\| < u\}$$

$$C = \{\omega \in \Omega : \|p(\omega) + q(\omega)\| < u\}.$$

Since the norm $\|\cdot\|$ satisfies the triangle inequality, it follows that $A \cap B \subset C$, so that

$$P(C) \geq P(A \cap B) \geq W(P(A), P(B)).$$

By (2.5.1), $P(A) = \nu_p(u), P(B) = \nu_q(v) \quad P(C) = \nu_{p+q}(t)$, so that

$$\nu_{p+q}(t) \geq W(\nu_p(u), \nu_q(v)),$$

and, hence

$$\nu_{p+q}(t) \geq \sup\{W(\nu_p(u), \nu_q(v)); u + v = t\} = \tau_W(\nu_p, \nu_q)(t).$$

As for property (N4), for every $t > 0$ and for every $p \in V$, one has from (2.5.1)

$$\begin{aligned}\tau_M(\nu_{\alpha p}, \nu_{(1-\alpha)p}(t)) &= \sup\{\nu_{\alpha p} \wedge \nu_{(1-\alpha\alpha)p}(t-u) : u \in [0,t]\} \\ &= \sup_{u \in [0,t]} \{P(\alpha\|p\| < u) \wedge P((1-\alpha)\|p\| < t-u)\} \\ &= P\left(\|p\| < \sup_{u \in [0,t]} \left\{\frac{u}{v} \wedge \frac{t-u}{1-\alpha}\right\}\right).\end{aligned}$$

Considering that

$$\frac{u}{v} \leq \frac{t-u}{1-\alpha} \quad \text{if, and only if,} \quad u \leq \alpha t,$$

one obtains, for every $t > 0$,

$$\tau_M(\nu_{\alpha p}, \nu_{(1-\alpha)p})(t) = \nu_p(t) \tag{2.5.2}$$

so that (S, ν, τ_W, τ_M) is a PPN space. Therefore, by virtue of Theorem 2.2.2, when the PN space (S, ν) is canonical, it is Šerstnev space under τ_W. □

It follows from (2.5.2) and Theorem 2.2.2 that a canonical EN space is a Šerstnev space.

The proof of the following theorem is immediate.

Theorem 2.5.2. *In an EN space (V, ν) let a relationship $\sim$ be defined via*

$$p \sim q \quad \textit{if, and only if,} \quad \nu_p = \nu_q.$$

Then $\sim$ is an equivalence relation on V. If $\overline{V} := V/\sim$ is the quotient space and

$$\tilde{\nu}_{\tilde{p}} := \nu_p \tag{2.5.3}$$

for every p in the equivalence class $\tilde{p}$, then $(\tilde{V}, \tilde{\nu})$ is a canonical EN space, called the quotient EN space of (V, ν).

2.6 Probabilistic Inner Product Spaces

Defining probabilistic inner product spaces (PIP spaces) has always seemed a natural step in the theory. However, attempts by Senechal (1965) and Fortuny (1984) encountered many obstacles and did not lead to a viable definition. The authors we have just mentioned gave their definition through a generalization of the Cauchy–Schwartz inequality. The definition we adopt was introduced in Alsina *et al.* (1997). For the definition one has to rely on d.f.s belonging to Δ rather than Δ^+ because the new notion should include that of classical inner product spaces and, of course, an inner product may take negative values; therefore, Δ provides a natural framework.

For every $F \in \Delta$ denote by $\overline{F}$ the d.f. in Δ defined, for every $t \in \mathbb{R}$, via

$$\overline{F}(t) := \ell^{-}(1 - F(t)). \tag{2.6.1}$$

Note that $\overline{\overline{F}} = F$ for every $F \in \Delta$ and that $F = \overline{F}$ if, and only if, F is symmetric.

Definition 2.6.1. Let V be a real linear space, $\mathcal{G}$a mapping from $V \times V$ into Δ, let τ and τ^* be multiplications on Δ, and let $G_{p,q}$ denote the value of $\mathcal{G}$ at the pair (p, q). Moreover, let the function $\nu : V \to \Delta^+$ be defined via

$$\nu_p(t) := \begin{cases} G_{p,q}(t^2), & \text{if } t > 0 \\ 0, & \text{if } t \leq 0. \end{cases} \tag{2.6.2}$$

Then the quadruple $(V, \mathcal{G}, \tau, \tau^*)$ is said to be a *Probabilistic Inner Product (PIP)* space if the following conditions hold, for all p, q, and r in V;

(P1a) $G_{p,q} \in \Delta^+$ for all $p \in V$ and $G_{\theta,\theta} = \epsilon_0$ where θ is the null vector in V;
(P1b) $G_{p,p} \neq \epsilon_0$ if $p \neq \theta$;
(P2) $G_{\theta,p} = \epsilon_0$;
(P3) $G_{p,q} = G_{q,p}$;
(P4) $G_{-p,q} = \bar{G}_{p,q}$;
(P5) $\nu_{p+q} \geq \tau(\nu_p, \nu_q)$;
(P6) $\forall \alpha \in [0, 1]$ $\nu_p \leq \tau^*(\nu_{\alpha p}, \nu_{(1-\alpha)p})$;
(P7) $\tau(G_{p,r}, G_{q,r}) \leq G_{p+q,r} \leq \tau^*(G_{p,r}, G_{q,r})$.

If $\tau = \tau_T$ and $\tau^* = \tau_{T^*}$ for some continuous t-norm T and its associated t-norm is T^* then $(V, \mathcal{G}, \tau, \tau^*)$ is a Menger PIP space, which will be denoted by $(V, \mathcal{G}, T)$. If $\tau^* = \tau_M$ and equality holds in (P6), then $(V, \mathcal{G}, \tau, \tau_M)$ is a Šerstnev PIP space. If (P1a) and (P2)–(P7) are satisfied, then $(V, \mathcal{G}, \tau, \tau^*)$ is a probabilistic pseudo-inner product space.

It is immediately seen that (V, ν, τ, τ^*) is a PN space and we shall refer to ν as the probabilistic norm derived from the probabilistic inner product ν. Notice again that $(V, \mathcal{G}, \tau, \tau_M)$ is a Šerstnev PIP space, in view of the fact that $\nu_{-p} = \nu_p$ (P6) may be replaced by

$$\forall \alpha, t \in \mathbb{R} \quad \nu_{\alpha p}(t) = \nu_p\left(\frac{t}{|\alpha|}\right). \tag{2.6.3}$$

If, for all p and q in V and for every $t \in \mathbb{R}$, one interprets the number $G_{p,q}(t)$ as "the probability that the inner product of p and q is less than t", then properties (P1)–(P4) are natural probabilistic versions of the corresponding properties of the real inner products. (P5) is the triangle inequality for the associated probabilistic norm. (P6) is a probabilistic version for the homogeneity property of a norm and is also needed in order to ensure that ν is indeed a probabilistic norm; finally, (P7) is a weak distributivity property that generalizes the usual bilinearity property of an inner product.

If $(V, \langle\cdot,\cdot\rangle)$ is a real inner product space, if τ is multiplication on Δ such that

$$\tau(\epsilon_a, \epsilon_b) = \epsilon_{a+b} \quad \text{for all a,b in } \mathbb{R},$$

and if $\mathcal{G} : V \times V \to \Delta$ is defined via

$$G_{p,q} := \epsilon_{a+b},$$

then $(V, \mathcal{G}, \tau, \tau^*)$ is a PIP space. Thus an ordinary metric and normed spaces, may, respectively, be viewed as special cases of PM and PN spaces, and a real inner product may be viewed as a special instance of PIP space.

The following definition introduces a class of PIP spaces that are particularly important because of probabilistic interpretation.

Definition 2.6.2. Let $(\Omega, \mathcal{A}, P)$ be a probability space, $(V, \langle\cdot,\cdot\rangle)$ a real inner product space and S a set of functions from Ω into V. Then $(S, \mathcal{G})$ is an EN space with base $(\Omega, \mathcal{A}, P)$ and target $(V, \langle\cdot,\cdot\rangle)$ if the following conditions hold:

(i) S is a real linear space under addition and scalar multiplication. The null element in S is a constant function $\theta(\omega) := \theta_V$ where θ_V is the null vector in V;

(ii) For all $p, q \in S$ and $t \in \mathbb{R}$ the set

$$\{\omega \in \Omega; \langle p(\omega), q(\omega)\rangle < t\}$$

belongs to $\mathcal{A}$, i.e. the composite function $\langle p, q\rangle$ from Ω into $\mathbb{R}$ defined by

$$\langle p, q\rangle(\omega) := \langle p(\omega), q(\omega)\rangle$$

is $\mathcal{A}$-measurable, or in other words, is a real random variable;

(iii) For all $p, q \in S$, $\mathcal{G}(p, q)$ is a d.f. of $\langle p, q\rangle$, i.e., for all $t \in \mathbb{R}$,

$$G_{p,q}(t) := P(\langle p, q\rangle < t) = P\{\omega \in \Omega : \langle p(\omega), q(\omega)\rangle < t\}. \qquad (2.6.4)$$

If $\langle p, q\rangle = 0$ almost everywhere (a.e.) for every $q \in S$, only if, $p = \theta$ then $(S, \mathcal{G})$ is a canonical EN space.

Theorem 2.6.1. *If $(S, \mathcal{G})$ is an EN space, then $(S, \mathcal{G}, \tau_W, \tau_{W^*})$ is a* pseudo-*PIP space, viz. $(S, \mathcal{G}, W)$ is a Menger PIP space.*

Proof. Properties (P1a), (P2), (P3), and (P4) are immediate, as is (P1b) when $(S, \mathcal{G})$ is canonical.

Next, it follows from Definition 2.6.2 that (S, ν) is an EN space. By Theorem 2.5.1, such a space is a pseudo-PN space in the sense of Šerstnev in which $\tau = \tau_W$. Condition (P5) is just the triangle inequality for this space; and since (2.6.3) holds, (2.2.5) yields (P6) with $\tau^* = \tau_M$, and since $\tau_M < \tau_{W^*}$ *a fortiori* with $\tau^* = \tau_M$.

It remains to establish (P7): using (2.6.4), one has, for all $p, q, r \in S$ and for every $t \in \mathbb{R}$,

$$G_{p+q,r}(t) = P(\langle p + q, r\rangle < t) = P(\langle p, r\rangle + \langle q, r\rangle < t).$$

Thus $G_{p+q,r}$ is the d.f. of the sum of the random variables $\langle p, r\rangle$ and $\langle q, r\rangle$. Let $C_{\langle p,r\rangle,\langle q,r\rangle}$ be the copula of these random variables, so that

$$C_{\langle p,r\rangle,\langle q,r\rangle}(G_{p,r}, G_{q,r})$$

is their joint d.f. Then, from Frank (1975), Moynihan *et al.* (1978) or Schweizer (1991), one has

$$G_{p+q,r} = \sigma_{C_{\langle p,r\rangle,\langle q,r\rangle}}(G_{p,r}, G_{q,r})$$

where, for every pair of d.f.s F and G and for every copula C

$$\sigma_C(F, G) := \int_{u+v<x} dC(F(u), G(v)).$$

Next one has, for every copula C and for every pair F and G of d.f.s

$$\tau_W(F, G) \le \sigma_C(F, G) \le \tau_{W^*}(F, G1).$$

This yields (P7), with $\tau = \tau_W$ and $\tau^* = \tau_{W^*}$ and concludes the proof. □

In particular, Theorem 2.6.1 applies to the product of random variables or of random vectors on a probability space $(\Omega, \mathcal{A}, P)$. In this case the set of random variables or vectors on $(\Omega, \mathcal{A})$, while the target is $\mathbb{R}^k (k \geq 1)$ endowed with the usual inner product

$$\langle x, y \rangle = \sum_{j=1}^{k} x_i y_j \quad (x, y \in \mathbb{R}^k).$$

The following is a simple but surprising result. Recall that a t-norm T is said to be *positive* if $T(a, b) > 0$, whenever $a > 0$ and $b > 0$.

Theorem 2.6.2. *Let $(V, \mathcal{G}, T)$ be a Menger PIP space and suppose that there is a pair of points $p, q \in V$ such that $G_{p,q}$ is strictly positive in $\mathbb{R}$, and $G_{p,p}$ and $G_{q,q}$ are both strictly positive on $]0, +\infty[$. Then T cannot be a positive t-norm.*

Proof. By (P7) and (P3) one has

$$G_{p+q,p+q} \geq \tau_T(G_{p+q,p}, G_{p+q,q}) \geq \tau_T(\tau_T(G_{p,p}, G_{p,q}), \tau_T(G_{p,q}, G_{q,q})).$$

Consequently, for all $t, u, v, w \in \mathbb{R}$ such that $t + u + v + w = 0$,

$$0 = G_{p+q,p+q}(0) \geq T(T(G_{p,p}(t), G_{p,q}(u), T(G_{p,q}(v), G_{q,q}(w)).$$ □

Lemma 2.6.1. *If $F \in \Delta$ is such that*

$$\tau_T(F, \bar{F}) = \epsilon_0, \tag{2.6.5}$$

where $\bar{F}$ is defined by (2.6.1), *then F is a proper d.f., i.e., F belongs to $\mathcal{D}$, or*

$$\lim_{t \to -\infty} F(t) = 0 \quad \textit{and} \quad \lim_{t \to +\infty} F(t) = 1.$$

Proof. When explicitly written, (2.6.5) reads

$$\sup_{u+v=t} T(F(u), \bar{F}(v)) := \begin{cases} 0, & \text{if } t \leq 0 \\ 1, & \text{if } t > 0. \end{cases}$$

Suppose that $\lim_{t \to +\infty} F(t) = a < 1$. Then, since $T(a, 1) = a$, one has, for all u and v in $\mathbb{R}$

$$T(F(u), \bar{F}(v)) \leq a,$$

whence (2.6.5) cannot hold.

Next, if $\lim_{t\to+\infty} F(t) = b > 0$, then

$$\lim_{t\to+\infty} \bar{F}(t) = \lim_{t\to+\infty}(1 - F(-t)) = 1 - \lim_{t\to-\infty} F(t) = 1 - b < 1,$$

and, again, (2.6.5) cannot hold. □

Lemma 2.6.2. *For a d.f. $F \in \Delta, \tau_M(F, \bar{F}) = \epsilon_0$ if, and only if, $F = \epsilon_c$ for some $c \in \mathbb{R}$.*

Proof. By Lemma 2.6.1, F is a proper d.f., $F \in \mathcal{D}$, whence $\bar{F}$ is also proper, $\bar{F} \in \mathcal{D}$. Then by duality, one has, for every $t \in]0, 1[$,

$$F^\wedge(t) + \bar{F}^\wedge(t) = 0. \tag{2.6.6}$$

But $\bar{F}^\wedge(t) = -F^\wedge(1 - t)$, so that (2.6.6) reads $F^\wedge(t) = F^\wedge(1 - t)$. Since $F^\wedge$ is non decreasing, this later equality can hold only if $F^\wedge$ is constant, say, equal to $c \in \mathbb{R}$. Hence $F = \epsilon_c$ and the lemma is proved. □

Theorem 2.6.3. *If $(V, \mathcal{G}, M)$ is a Menger PIP space, then it is a real inner product space. i.e., there exists a real inner product $\langle\cdot,\cdot\rangle : V \times V \to \mathbb{R}$ such that, for all $p, q \in V$,*

$$G_{p,q} = \epsilon_{\langle p,q\rangle}.$$

Proof. Since $\tau_M = \tau_{M^*}$, it follows at once from (P7) that, for all $p, q, r \in V$

$$G_{p+q,r} = \tau_M(G_{p,r}, G_{q,r}).$$

Letting $q = -p$ and using (P2) and (P4) yields

$$\epsilon_0 = \tau_M(G_{p,r}, \bar{G}_{p,r}).$$

Thus, by Lemma 2.6.2 there exists a real number c such that $G^\wedge_{p,r} = \epsilon_0$. Now, define a mapping $\langle\cdot,\cdot\rangle : V \times V \to \mathbb{R}$ through

$$\langle p, q\rangle := G^\wedge_{p,q}$$

for all $p, q \in V$. Notice that, by duality, one has

$$\langle p + q, r\rangle = G^\wedge_{p+q,r} = [\tau_M(G_{p,r}, G_{q,r})]^\wedge = G^\wedge_{p,q} + G^\wedge_{q,r} = \langle p, r\rangle + \langle q, r\rangle.$$

□

It is known (see Corollary 8.2.2 and 8.2.3 in Schweizer and Sklar (1983)) that, if $(V, \mathcal{F}, M)$ is a Menger PM space, then

$$F^\wedge_{p,r}(c) \le F^\wedge_{p,q}(c) + F^\wedge_{q,r}(c)$$

for all $p, q, r \in V$. Similarly, if (V, ν, M) is a Menger PN space, then each of the functions ν_c defined on V via

$$\nu_c(p) := \nu^\wedge_p(c)$$

is a pseudo-norm on V. It is therefore no surprise that many results, mainly in the area of fixed points, that hold in ordinary metric or normed spaces extend at once to Menger spaces under M.

2.7 Open Questions

(1) Let $(S, \mathcal{G})$ be an EN space with base $(\Omega, \mathcal{A}, P)$ and target $(V, \langle\cdot,\cdot\rangle)$. Then for every $\omega \in \Omega$, the function $i_\omega : S \times S \to \mathbb{R}$ defined by

$$i_\omega\langle p, q\rangle := \langle p(\omega), q(\omega)\rangle$$

is a pseudo-inner product on S. Since distinct functions p and q may agree at a particular point $\omega \in \Omega$, so that $p(\omega) = q(\omega)$ while $p \neq q$, i_ω need not be an inner product on S. Now, noting that

$$G_{p,q}(t) = P(i_\omega < t)$$

is the P-probability of the set of all pseudo-inner products i_ω for which the inner product of p and q is less than t, the EN space $(S, \mathcal{G})$ is a *pseudo-inner product generated space,* which is generated by the collection $\{i_\omega : \omega \in \Omega\}$.

(2) Is there a good analogue of the Cauchy–Schwartz inequality in PIP space?

(3) The question of orthogonality in PN spaces is entirely open. Very likely this is a deep problem, for, even in the "classical" case, important developments arose (see the papers by P. Jordan and J. von Neumann (1935) and J.M. Fortuny (1984).

(4) Which PN space can be derived from a PIP space? In other words, is there an analogue of the Jordan–Von Neumann theorem (1995) that characterizes those normed spaces that derive from an inner product space?

Chapter 3

The Topology of PN Spaces

In the following sections we shall establish the fundamental facts about the topology of PN spaces.

3.1 The Topology of a PN Space

Note that, given a PN space (V, ν, τ, τ^*), a probabilistic metric $\mathcal{F} : V \times V \to \Delta^+$ is defined via

$$\mathcal{F}(p, q) = F_{pq} := \nu_{p-q},$$

so that the triple $(V, \mathcal{F}, \tau)$ becomes a PM space. Since the triangle function τ is continuous, the system of neighborhoods

$$\mathcal{N} := \cup_{p \in V} \cup_{t>0} \{N_p(t) : p \in V\},$$

called the *strong neighborhood system,* where

$$N_p(t) := \{q \in V : d_S(F_{pq}, \varepsilon_0) < t\} = \{q \in V : \nu_{p-q}(t) > 1 - t\},$$

determines a topology on V, called the *strong topology* (see Theorem 12.1.2 in Schweizer and Sklar (1983) and Sempi (2004)). We shall state this fact as a theorem.

Theorem 3.1.1. *Let (V, ν, τ, τ^*) be a PN space with τ continuous. Then the strong neighborhood system $\mathcal{N}$ determines a Hausdorff topology for V.*

Later on, we shall consider the following fundamental concepts regarding the strong neighborhoods which generate the strong topology of a PN space.

Definition 3.1.1. Let (V, ν, τ, τ^*) be a PN space, then

(i) A sequence (p_n) in V is said to be strongly convergent to a point p in V, and we write $p_n \to p$ or $\lim p_n = p$, if for each $\lambda > 0$, there exists a positive integer m such that $p_n \in N_p(\lambda)$, for $n \geq m$;
(ii) A sequence (p_n) in V is called a strong Cauchy sequence if for every $\lambda > 0$ there is a positive integer N such that $\nu_{p_n - p_m}(\lambda) > 1 - \lambda$, whenever $m, n > N$;
(iii) The PN space (V, ν, τ, τ^*) is said to be distributionally compact ($\mathcal{D}$-compact) if every sequence (p_n) in V has a convergent subsequence (p_{n_k}). A subset A of a PN space (V, ν, τ, τ^*) is said to be $\mathcal{D}$-compact if every sequence (p_m) in A has a subsequence (p_{m_k}) that converges to a point $p \in A$;
(iv) In the strong topology, the closure $\overline{N_p(\lambda)}$ of $N_p(\lambda)$ is defined by

$$\overline{N_p(\lambda)} := N_p(\lambda) \cup N_p(\lambda)',$$

where $N_p(\lambda)'$ is the set of limit points of all convergent sequences in $N_p(\lambda)$.

Since the strong topology is first-countable and Hausdorff, it can be completely specified in terms of the strong convergence of sequences.

3.2 The Uniform Continuity of the Probabilistic Norm

In a "classical" normed space $(V, \|\cdot\|)$ the norm is a uniformly continuous mapping from V into $\mathbb{R}_+$. This fact is an immediate consequence of the inequality

$$\|x + y\| \leq \|x\| + \|y\|, \tag{3.2.1}$$

which holds for all x and y in V. Assume $\|x_n - x\| \to 0$ and $\|y_n - y\| \to 0$ as n goes to $+\infty$, then applying (3.2.1) one obtains for every $n \in \mathbf{N}$

$$\|x_n - y_n\| \leq \|x_n - x\| + \|x - y\| + \|y - y_n\|$$

and

$$\|x - y\| \leq \|x - x_n\| + \|x_n - y_n\| + \|y_n - y\|.$$

Together these two inequalities yield, for every $n \in \mathbf{N}$,

$$|\|x - y\| - \|x_n - y_n\|| \leq \|x - x_n\| + \|y - y_n\|,$$

from which the assertion follows immediately.

The proof of the analogous result for a PN space must avoid the use of subtraction, which is not allowed in the semigroup (Δ^+, τ) and is similar, in fact simpler, than the corresponding proof of the uniform continuity of the distance function in a PM space (see Section 12.2 in Schweizer and Sklar (1983)). It also relies on Lemma 12.2.1 from Schweizer and Sklar (1983), which we reproduce below.

Lemma 3.2.1. *Let τ be a continuous triangle function and S the set of all distance d.f.s F, G, and H such that*

$$F \geq \tau(H, G) \quad \textit{and} \quad G \geq \tau(H, F). \tag{3.2.2}$$

Then, for every $\eta > 0$ there exists $\delta = \delta(\eta) > 0$ such that, if (F, G, H) is in S and if $d_S(H, \varepsilon_0) < \delta$, then $d_S(F, G) < \eta$.

We can now state and prove the announced result.

Theorem 3.2.1. *Let (V, ν, τ, τ^*) be a PN space with τ continuous. If V is endowed with the strong topology and Δ^+ with the topology of Sibley's metric d_S, then the probabilistic norm $\nu : V \to \Delta^+$ is uniformly continuous.*

Proof. Let p and q be in V. Then

$$\nu_p \geq \tau(\nu_{p-q}, \nu_q) \quad \text{and} \quad \nu_q \geq \tau(\nu_{p-q}, \nu_p)$$

so that (3.2.2) is satisfied, with $F = \nu_p, G = \nu_q$ and $H = \nu_{p-q}$. Therefore, for every $\eta > 0$ there is $\delta = \delta(\eta)$ such that $d_S(\nu_p, \nu_q) < \eta$ whenever $d_S(\nu_{p-q}, \varepsilon_0) < \delta$, i.e. whenever q belongs to the neighborhood $N_p(\delta)$ of p. □

Notice that we have chosen to prove Theorem 3.2.1 directly; it might have been proved as a consequence of Theorem 12.2.2 in Schweizer and Sklar (1983) by putting $\nu_p = F_{p,\theta}$ for every $p \in V$. Notice also that Theorem 3.2.1 holds without any change for a Šerstnev space (V, ν, τ) with τ continuous since the triangle function τ^* plays no role in the proof.

3.3 A PN Space as a Topological Vector Space

One still has to investigate whether, or rather, when a PN space (V, ν, τ, τ^*) is a *topological vector space* (TV space). To this end the interplay between topology and the definition of a PN space has to be studied. This was accomplished in Alsina, Schweizer, and Sklar (1977). For a PN space to be a TV space, the sum $+$ on V, for every $\alpha \in \mathbb{R}$; the mapping from V into V

defined by $p \mapsto \alpha p$; and, for every $p \in V$, the mapping from $\mathbb{R}$ into V defined by $\alpha \mapsto \alpha p$ must be continuous. We shall study the three mappings just mentioned separately.

Theorem 3.3.1. *Let (V, ν, τ, τ^*) be a PN space with τ continuous. Let V be endowed with the strong topology, let Δ^+ be endowed with the topology of Sibley's metric d_S, and $V \times V$ with the corresponding product topology. Then the sum $+$ on V is a jointly uniformly continuous map from $V \times V$ into V, so that $(V, +)$ is a topological group.*

Proof. For a fixed $\eta > 0$ let p, p', q, q' in V be such that

$$d_S(\nu_{p-p'}, \varepsilon_0) > \eta \quad \text{and} \quad d_S(\nu_{q-q'}, \varepsilon_0) > \eta.$$

Then

$$\nu_{(p+q)-(p'+q')} = \nu_{(p-p')+(q-q')} \geq \tau(\nu_{p-p'}, \nu_{q-q'})$$

so that Lemma 1.3.3 yields

$$d_S(\nu_{(p+q)-(p'+q')}, \varepsilon_0) \leq d_S(\tau(\nu_{p-p'}, \nu_{q-q'}, \varepsilon_0),$$

from which the assertion follows since τ is uniformly continuous. □

Combining Theorems 3.2.1 and 3.3.1, *viz.* considering the composition of ν and $+$, one obtains:

Corollary 3.3.1. *Under the assumptions of Theorem 3.2.1 the mapping $s : V \times V \to \Delta^+$ defined by $s(p, q) := \nu_{p+q}$ is uniformly continuous.*

In order to prove the continuity of the mapping $p \to \alpha p \, (\alpha \in \mathbb{R})$ we shall need the following two lemmas.

Lemma 3.3.1. *Let (V, ν, τ, τ^*) be a PN space and let α and β be two real numbers with $0 \leq \alpha \leq \beta$. Then, for all $p \in V$, one has*

$$\nu_{\beta p} \leq \nu_{\alpha p}. \tag{3.3.1}$$

Proof. There is $\lambda \in [0, 1]$ such that $\alpha = \lambda\beta$. Then

$$\nu_{\beta p} = \nu_{\lambda\beta p + (1-\lambda)\beta p} \leq \tau^*(\nu_{\lambda\beta p}, \nu_{(1-\lambda)\beta p}) \leq \tau^*(\nu_{\lambda\beta p}, \varepsilon_0) = \nu_{\lambda\beta p} = \nu_{\alpha p}.$$

□

Lemma 3.3.2. *For all $\alpha \in \mathbb{R}, p \in V$ and for every $\eta > 0$ there is $\delta = \delta(\eta) > 0$ such that, if $d_S(\nu_p, \varepsilon_0) < \delta$, then $d_S(\nu_{\alpha p}, \varepsilon_0) < \eta$.*

Proof. Because of (N2) one may assume, without loss of generality, that $\alpha \geq 0$. If $\alpha \in [0,1]$, Lemma 3.3.1 implies $d_S(\nu_{\alpha p}, \varepsilon_0) \leq d_S(\nu_p, \varepsilon_0)$, whence the assertion. For $\alpha > 1$ there exists $k \in \mathbf{N}$ such that $k-1 \leq \alpha < k$; then $\nu_{\alpha p} \geq \nu_{kp}$ and the repeated use of (N3) gives

$$\begin{aligned}\nu_{kp} &\geq \tau(\nu_{(k-1)p}, \nu_p) \geq \tau(\tau(\nu_{(k-2)p}, \nu_p), \nu_p)\\ &= \tau^2(\nu_{(k-2)p}, \nu_p, \nu_p) = \cdots = \tau^k(\nu_p, \nu_p, \ldots, \nu_p).\end{aligned}$$

where τ^k is the k-th serial iterate of τ. Therefore,

$$d_S(\nu_{kp}, \varepsilon_0) \leq d_S(\tau^k(\nu_p, \nu_p, \ldots, \nu_p), \varepsilon_0)).$$

But τ^k is jointly uniformly continuous, so that for every $\eta > 0$ there exists $\delta + \delta(\eta) > 0$ such that $d_S(\tau^k(\nu_p, \nu_p, \ldots, \nu_p), \varepsilon_0)) < \eta$ whenever $d_S(\nu_{kp}, \varepsilon_0) < \delta$, which concludes the proof. □

Theorem 3.3.2. *For every $\alpha \in \mathbb{R}$, the mapping $p \mapsto \alpha p$ is uniformly continuous on V.*

Proof. Given $\eta > 0$, it suffices to replace p in Lemma 3.3.2 by $p - q$ in order to have

$$d_S(\nu_{\alpha(p-q)}, \varepsilon_0) < \eta \quad \text{whenever } d_S(\nu_{p-q}, \varepsilon_0) < \delta$$

which concludes the proof. □

In general, the mapping $\mathbb{R} \ni \alpha \mapsto \alpha p$ ($p \in V$) is not continuous, as the following example shows.

Example 3.3.1. Take $V = \mathbb{R}, \tau = \tau_W, \tau^* = \tau_M$ and define the probabilistic norm ν via $\nu_0 = \varepsilon_0$ and for $p \neq 0$, via

$$\nu_p = \frac{1}{|p|+2}\varepsilon_0 + \frac{|p|+1}{|p|+2}\varepsilon_\infty.$$

Then $(\mathbb{R}, \nu, \tau_W, \tau_M)$ is a PN space; but, for every $p \neq 0$ and for every sequence $\{\alpha_n\}$ that tends to 0, the sequence $\{\nu_{\alpha_n p}\}$ converges weakly to $(\varepsilon_0 + \varepsilon_\infty)/2$ rather than ε_0. Therefore, $\alpha \mapsto \alpha p$ is not continuous.

Theorem 3.3.3 below will provide a *sufficient* condition that ensures that the continuity of the mapping $\alpha \mapsto \alpha p$ for all $p \in V$ (see Theorem 4 in Alsina, Schweizer, and Sklar (1997)).

Lemma 3.3.3. *If the triangle function τ^* is Archimedean then, for every $p \in V$ such that $\nu_p \neq \varepsilon_\infty$ and for every $\eta > 0$, there exists $\delta > 0$, such that*

$$d_S(\nu_{\alpha p}, \epsilon_0) < \eta \quad \textit{whenever } \|\alpha\| < \delta,$$

in other words

$$\lim_{\alpha \to 0} d_S(\nu_{\alpha p}, \varepsilon_0) = 0.$$

Proof. In view of Lemma 3.3.1 and property (N2) it suffices to show that $\{\nu_{p_n}\}$ converges weakly to ε_0 as n goes to $+\infty$, where p belongs to V and $p_n := p/2^n (n \in \mathbb{N})$. It follows from (N4) that

$$\varepsilon_\infty < \nu_{p_n} \leq \tau^* \left(\nu_{p_{n+1}}, \nu_{p_{n+1}}\right) \leq \nu_{p_{n+1}};$$

therefore, the sequence $\{\nu_{p_n}\}$ is increasing. Let F be its weak limit. Since τ^* is continuous it follows that $\tau^*(F, F) = F$, so that F is an idempotent of τ^* with $F > \varepsilon_\infty$. But τ^* is Archimedean; therefore $F = \epsilon_0$. □

Theorem 3.3.3. *Let (V, ν, τ, τ^*) be a PN space in which τ^* is Archimedean and $\nu_p \neq \epsilon_\infty$ for all $p \in V$. Then, for every $p \in V$, the mapping $\mathbb{R} \ni \alpha \mapsto \alpha p$ is uniformly continuous.*

Proof. For every $p \in V$ and for all $\alpha, \beta \in \mathbb{R}$ one has

$$\nu_{\alpha p} \geq \tau(\nu_{\beta p}, \nu_{(\alpha-\beta)p}) \quad \text{and} \quad \nu_{\beta p} \geq \tau(\nu_{\alpha p}, \nu_{(\alpha-\beta)p}).$$

It follows from Lemma 3.2.1 that, for every $\eta > 0$, there exists $\delta = \delta(\eta) > 0$ such that

$$d_S(\nu_{(\alpha-\beta)p, \varepsilon_0}) < \eta,$$

if $d_S(\nu_{\alpha p}, \nu_{\beta p}) < \delta$. □

It is now possible to collect the results of this section in the following theorem, which provides a *sufficient* condition for a PN space to be a topological vector space.

Theorem 3.3.4. *Every PN space (V, ν, τ, τ^*) in which the triangle function τ^* is Archimedean is a topological vector space.*

It is not hard to provide the examples of a PN space that is not a topological vector space; the examples that follow are taken from Lafuerza-Guillén (2001).

Example 3.3.2. Let $(X, \|\cdot\|)$ be a normed space and define on it a probabilistic norm $\nu : V \to \Delta^+$ via $\nu_\theta = \varepsilon_0$ and, if $p \neq \theta$,

$$\nu_p := \varepsilon_{\frac{\beta+\|p\|}{\beta}} \quad (\beta > 0).$$

Further, let τ be a triangle function such that, for all c and d with $c > 0$ and $d > 0$,

$$\tau(\varepsilon_c, \varepsilon_d) \leq \varepsilon_{c+d}. \tag{3.3.2}$$

Then (V, ν, τ, Π_M) is a PN space that is neither a topological vector space nor a Šerstnev space.

Properties (N1) and (N2) are obvious. As for (N3), one has

$$\begin{aligned}\nu_{p+q} &= \varepsilon_{\frac{\beta+\|p+q\|}{\beta}} \geq \varepsilon_{\frac{\beta+\|p\|+\|q\|}{\beta}} \\ &\geq \varepsilon_{\frac{\beta+\|p\|}{\beta}+\frac{\beta+\|q\|}{\beta}} \geq \tau\left(\varepsilon_{\frac{\beta+\|p\|}{\beta}}, \varepsilon_{\frac{\beta+\|q\|}{\beta}}\right) = \tau(\nu_p, \nu_q).\end{aligned}$$

For every $\alpha \in]0,1[$ and for every $t > 0$,

$$Min\{\varepsilon_{\frac{\beta+\alpha\|p\|}{\beta}}(t), \varepsilon_{\frac{\beta+(1-\alpha)\|p\|}{\beta}}(t)\} \geq \varepsilon_{\frac{\beta+\|p\|}{\beta}}(t) = \nu_p(t).$$

Thus (V, ν, τ, Π_M) is a PN space; however, it cannot be a topological vector space. In fact let $\{\alpha_n\}$ be a sequence of real numbers such that $\lim_{n\to+\infty} \alpha_n = 0$. Then, for every $p \neq \theta$

$$\lim_{n\to+\infty} \nu_{\alpha_n p} = \lim_{n\to+\infty} \varepsilon_{\frac{\beta+\alpha_n\|p\|}{\beta}} = \varepsilon_1 \neq \varepsilon_0.$$

Moreover, it follows from Lemma 7.2.13 in Schweizer and Sklar (1983) that

$$\begin{aligned}\tau_M(\nu_{\alpha p}, \nu_{(1-\alpha)p}) &= \tau_M\left(\varepsilon_{\frac{\beta+\alpha\|p\|}{\beta}}, \varepsilon_{\frac{\beta+(1-\alpha)\|p\|}{\beta}}\right) \\ &= \varepsilon_{\frac{2\beta+\|p\|}{\beta}} \leq \varepsilon_{\frac{\beta+\|p\|}{\beta}} = \nu_p.\end{aligned}$$

As a consequence, for every $t \in]1 + \frac{\|p\|}{\beta}, 2 + \frac{\|p\|}{\beta}[$, one has

$$\tau_M(\nu_{\alpha p}, \nu_{(1-\alpha)p}) < \nu_p(t),$$

so that the space considered is not a Šerstnev space.

Example 3.3.3. As in the previous example let $(X, \|\cdot\|)$ be a normed space, τ a triangle function that satisfies condition (3.3.2), and define the probabilistic norm through

$$\nu_p := \varepsilon_{\frac{\|p\|}{\beta+\|p\|}} \quad (p \in V, \beta > 0).$$

Then the quadruple (V, ν, τ, Π_M) is a PN space and a topological vector space, but it is not a Šerstnev space.

Properties (N1) and (N2) are obvious. As for (N3), the function

$$]0,+\infty[\ni t \mapsto \frac{t}{\beta+t}$$

is non-decreasing. Therefore,

$$\nu_{p+q} = \varepsilon_{\frac{\|p+q\|}{\beta+\|p+q\|}} \geq \varepsilon_{\frac{\|p\|+\|q\|}{\beta+\|p\|+\|q\|}} \geq \tau\left(\varepsilon_{\frac{\|p\|}{\beta+\|p\|+\|q\|}}, \varepsilon_{\frac{\|q\|}{\beta+\|p\|+\|q\|}}\right) \tag{3.3.3}$$

$$\geq \tau\left(\varepsilon_{\frac{\|p\|}{\beta+\|p\|}}, \varepsilon_{\frac{\|q\|}{\beta+\|q\|}}\right) = \tau(\nu_p, \nu_q). \tag{3.3.4}$$

Thus (V, ν, τ, Π_M) is a PN space. Now, let $\{\alpha_n : n \in \mathbb{N}\}$ be a sequence of real numbers such that $\lim_{n\to+\infty} \alpha_n = 0$; then, for every $p \in V$, one has

$$\lim_{n\to+\infty} \nu_{\alpha_n p} = \lim_{n\to+\infty} \varepsilon_{\frac{\alpha_n\|p\|}{\beta+\alpha_n\|p\|}} = \varepsilon_0,$$

which implies that the PN space under study is a topological vector space.

Choose arbitrarily $p \in V$ and $\alpha \in]0,1[$; Lemma 7.2.13 in Schweizer and Sklar (1983) yields

$$\tau_M(\nu_{\alpha p}, \nu_{(1-\alpha)p}) = \varepsilon_{\frac{\beta\|p\|+2\alpha(1-\alpha)\|p\|^2}{\beta^2+\beta\|p\|+\alpha(1-\alpha)\|p\|^2}}.$$

For every t belonging to the interval

$$\left]\frac{\|p\|}{\beta+\|p\|}, \frac{\beta\|p\|+2\alpha(1-\alpha)\|p\|^2}{\beta^2+\beta\|p\|+\alpha(1-\alpha)\|p\|^2}\right[$$

one has

$$\nu_p(t) = 1 > 0 = \tau_M(\nu_{\alpha p}, \nu_{(1-\alpha)p})(t)$$

so that (V, ν, τ, Π_M) is not a Šerstnev space.

3.4 Completion of PN Spaces

The completion of PM spaces was studied by Muśtari (1967), Sherwood (1966, 1971) and Sempi (1992). They proved that every PM space admits a completion. The question naturally arises of whether or not a PN space admits a completion. This question is answered in the positive both for a Šerstnev space (V, ν, τ) and for a general PN space (V, ν, τ, τ^*). We shall deal separately with these two cases. For the sake of completeness, it is expedient to repeat that part of the proof necessary to establish the completeness of a PM space; this is reproduced from Sherwood (1966). Sherwood's proof is similar, in principle, to the standard argument used to prove that every metric space has a completion, which is unique up to isometries; the main difference is that while $(\mathbb{R}, +)$ is a group, where as a consequence subtraction is possible, $(\mathcal{D}^+, \tau)$ is only a semigroup where no subtraction is defined. Of course extra care has to be taken because one has to take into account the linear structure of the space V (Lafuerza-Guillén, Rodríguez Lallena, & Sempi, 1995).

Theorem 3.4.1. *Every Šerstnev space (V, ν, τ) with a continuous triangle function τ has a completion, viz. it is isometric to dense linear subspace of a complete Šerstnev space (V', ν', τ). This completion is unique up to isometries.*

Proof. We recall that the topology of the PN space (V, ν, τ) is that of the corresponding PM space $(V, \mathcal{F}, \tau)$ where, for all $p, q \in V, \mathcal{F}(p, q) = F_{pq} := \nu_{p-q}$. Two Cauchy sequences $\{p_n\}$ and $\{q_n\}$ of elements of V are said to be equivalent $\{p_n\} \sim \{q_n\}$ if F_{p_n, q_n} converges weakly to ε_0 when $n \to +\infty$, or, equivalently, when

$$d_S\left(F_{p_n \cdot q_n}, \varepsilon_0\right) \to 0 \quad \text{as } n \to +\infty.$$

The symmetry and reflexivity of $\sim$ are obvious. As for transitivity, assume that $\{p_n\}$, $\{q_n\}$, and $\{r_n\}$ are Cauchy sequences of elements of V with $\{p_n\} \sim \{q_n\}$ and $\{q_n\} \sim \{r_n\}$. It follows from (PM4) that, for every $n \in \mathbb{N}$,

$$F_{p_n, r_n} \geq \tau(F_{p_n, q_n}, F_{q_n, r_n})$$

whence

$$0 \leq d_S(F_{p_n, r_n}, \varepsilon_0) \leq d_S(\tau\left(F_{p_n, q_n}, F_{q_n, r_n}, \varepsilon_0\right)).$$

Both d_S and τ are continuous so that F_{p_n, r_n} tends to ε_0. Thus $\sim$ is an equivalence relation on the set of all Cauchy sequences of elements of V. Denote by V' the family of equivalence classes.

Let $p', q' \in V'$ and let $\{p_n\}$ and $\{q_n\}$ belong to p' and q' respectively. Since $\{p_n\}$ and $\{q_n\}$ are Cauchy sequences, for every $\eta > 0$ there exists a natural number $n_0 = n_0(\eta)$ such that, for all $n, m \geq n_0$ both

$$d_S(F_{p_n, p_m}, \varepsilon_0) < \eta \quad \text{and} \quad d_S(F_{q_n, q_m}, \varepsilon_0) < \eta$$

hold. On the other hand τ is uniformly continuous; therefore, for every $\delta > 0$ there exists $\eta = \eta(\delta) > 0$ such that, for every $F \in \Delta^+$,

$$d_S(\tau(F, G), \tau(F, H) \left\langle \frac{\delta}{2} \right\rangle, \tag{3.4.1}$$

whenever $d_S(G, H) < \eta$. Thus, since

$$F_{p_n,q_n} \geq \tau(F_{p_n,q_m}, F_{q_m,p_m}),$$

setting $H = \varepsilon_0, F = F_{p_n,q_m}$, and $G = F_{q_m,q_n}$ in (3.4.1) one has, for all $n, m \geq n_0$ and for all $t > 0$,

$$F_{p_n,q_n}(t + \delta) \geq \tau(F_{p_n,q_m}, F_{q_m,q_n})(t + \delta) \geq F_{p_n,q_m}\left(t + \frac{\delta}{2}\right) - \frac{\delta}{2};$$

using (3.4.1) again with $H = \varepsilon_0, F = F_{p_m,q_n}$, and $G = F_{p_n,p_m}$ yields

$$F_{p_n,q_n}(t + \delta) \geq \tau(F_{p_n,p_m}, F_{p_m,q_m})\left(t + \frac{\delta}{2}\right) - \frac{\delta}{2} \geq F_{p_m,q_m}(t) - \delta.$$

In the same way one proves that $F_{p_n,q_n}(t - \delta) - \delta \leq F_{p_m,q_m}(t)$. Combining the last two inequalities yields

$$F_{p_n,q_n}(t - \delta) - \delta \leq F_{p_m,q_m}(t) \leq F_{p_n,q_n}(t + \delta) + \delta,$$

i.e. $d_S(F_{p_n,q_n}, F_{p_m,q_m}) < \delta$ whenever $n, m \geq n_0$. Therefore $\{V\}$ is a Cauchy sequence in the complete space (Δ^+, d_S) which is complete; thus there exists a d.f. $F' \in \Delta^+$ such that F_{p_n,q_n} converges weakly to F'. Then set

$$F'(p', q') := F'.$$

The d.f. F' is independent of the particular representatives chosen from p' and q'. If $\{p_n^*\}$ and $\{q_n^*\}$ are two other sequences in p' and q' respectively, then

$$F_{p_n^*,q_n^*} \geq \tau(F_{p_n^*,p_n}, F_{p_n,q_n^*}) \geq \tau^2[F_{p_n^8,p_n}, F_{p_n,q_n}, F_{q_n,q_n^*}]$$

so that

$$\lim_{n\to+\infty} F_{p_n^*,q_n^*} = \lim_{n\to+\infty} F_{p_n,q_n}.$$

Now V' is the set of equivalence classes of Cauchy sequences of elements of V. In order to prove that V' is a linear space, let p' and q' belong to V' and let $\{p_n\}$ and $\{q_n\}$ be Cauchy sequences of elements of V with $\{p_n\} \in p'$ and $\{q_n\} \in q'$. Since V is a linear space, one has, for every $n \in \mathbb{N}, p_n + q_n \in V$. It is possible to define a sum of p' and q' in such a manner that $p' + q'$ belongs to V'. Since (V, F, τ) with $F(p, q) := \nu_{p-q}$ is a PM space, for n and m in $\mathbb{N}$, one has

$$\begin{aligned} F_{p_n+q_n,p_m+q_m} &= \nu_{(p_n+q_n)-(p_m+q_m)} = \nu_{(p_n-q_n)+(p_m-q_m)} \\ &\leq \tau(\nu_{p_n-q_n}, \nu_{p_m-q_m}). \end{aligned}$$

Taking into account Lemma 1.3.3 one has

$$\begin{aligned} d_S(F_{p_n+q_n,p_m+q_m},\varepsilon_0) &\le d_S(\tau(\nu_{p_n-q_n},\nu_{p_m-q_m}),\varepsilon_0) \\ &= d_S(\tau(F_{p_n,q_n},F_{p_m,q_m}),\varepsilon_0). \end{aligned}$$

The continuity of both d_S and τ ensures that $F_{p_n+q_n,p_m+q_m}$ converges weakly to ε_0 when both $n,m \to +\infty$. Therefore $\{p_n+q_n\}$ is a Cauchy sequence in V, and as a consequence it belongs to an element of V', which will be denoted by τ'. Then define $p'+q' := r'$; this is a good definition, since it does not depend on the elements of p' and q' selected, for, if $\{p_n\}$ and $\{p_n^*\}$ both belong to p' and $\{q_n\}$ and $\{q_n^*\}$ both belong to q', then

$$\begin{aligned} F_{p_n+q_n,p_n^*+q_n^*} &= \nu_{(p_n+q_n)-(p_n^*+q_n^*)} = \nu_{(p_n-p_n^*)+(q_n-q_n^*)} \\ &\ge \tau(\nu_{p_n-p_n^*},\nu_{q_n-q_n^*}) = \tau(F_{p_n-p_n^*},F_{q_n-q_n^*}) \end{aligned}$$

so that

$$d_S(F_{p_n+q_n,p_n^*+q_n^*},\varepsilon_0) \le d_S(\tau\left(F_{p_n,p_n^*},F_{q_n,q_n^*}\right),\varepsilon_0).$$

Since both d_S and τ are continuous, the sequence $\{F_{p_n+q_n,p_n^*+q_n^*}\}$ converges weakly to ε_0, in other words, $\{p_n+q_n\} \sim \{p_n^*+q_n^*\}$. Thus the sum defined above is a good definition, which immediately satisfies the properties of an abelian group.

For every $\alpha \in \mathbb{R}$, and for every Cauchy sequence $\{p_n\}$ of elements of V, $\{\alpha p_n\}$ is also a Cauchy sequence in V. This is obvious if $\alpha = 0$. If $\alpha \ne 0$, one has, for every $t > 0$,

$$F_{\alpha p_n,\alpha p_m}(t) = \nu_{\alpha p_n-\alpha p_m}(t) = \nu_{p_n-p_m}\left(\frac{t}{|\alpha|}\right) = F_{p_n-p_m}\left(\frac{t}{|\alpha|}\right),$$

which tends to 1 for every $t > 0$, when $n,m \to +\infty$. In other words, $F_{\alpha p_n,\alpha p_m}$ converges weakly to ε_0. Thus $\{\alpha p_n\}$ is a Cauchy sequence in V: denote by u' the element of V' to which it belongs and define $\alpha p' := u'$. Again, this is a good definition: in fact, let $\{p_n\}$ and $\{p_n^*\}$ belong to p'. Then one has, for every $t > 0$,

$$F_{\alpha p_n,\alpha p_n^*}(t) = \nu_{\alpha(p_n-p_m^*)}(t) = \nu_{p_n-p_n^*}\left(\frac{t}{|\alpha|}\right)$$

which tends to 1 for every $t > 0$, when $n \to +\infty$, whence $\{\alpha p_n\} \sim \{\alpha p_n^*\}$. Therefore V' is a (real) linear space. All that remains to show is that the distance d.f. $\mathcal{F}'$ derives from a probabilistic norm ν' on V'. To this end, let p' be any element in V', choose $\{p_n\} \in p$ with $p_n \in V$ for every $n \in \mathbf{N}$ and define

$$\nu'_{p'} := F'(p',\theta') = \lim_{n\to+\infty} F_{p_n,\theta} = \lim_{n\to+\infty} \nu_{p_n}. \tag{3.4.2}$$

Thus, for p' and q' in V', one has

$$F'_{p',q'} + \lim_{n\to+\infty} F_{p_n,q_n} = \lim_{n\to+\infty} \nu_{p_n-p_n} = \nu'_{p'-q'}.$$

It is now an easy task to check that the triplet (V', ν', τ) is a Šerstnev PN space. In fact, the sequence $\{o_n\}$ where for every $n \in \mathbf{N}, o_n = \theta$ belongs to the origin $\theta' \in V'$. Therefore

$$\nu_{\theta'} = \lim_{n\to+\infty} \nu_{o_n} = \nu_\theta.$$

If $\nu'_{p'} = \varepsilon_0$ and if $\{p_n\}$ belongs to p', one has

$$\lim_{n\to+\infty} \nu_{p_n} = \varepsilon_0$$

so that $\{p_n\}$ belongs to θ'. Condition (N1) is thus satisfied.

Let p' and q' in V', and let $\{p_n\}$ and $\{q_n\}$ belong to p' and q' respectively. Then, on account of the continuity of τ, one has

$$\nu'_{p'+q'} = \lim_{n\to+\infty} \nu_{p_n+q_n} \geq \lim_{n\to+\infty} \tau(\nu_{p_n}, \nu_{q_n}) = \tau(\nu'_{p_n}, \nu'_{q_n}),$$

which means that condition (N3) is also fulfilled. For any $p' \in V'$ let $\{p_n\}$ belong to p'. For every $\alpha \neq 0$ and for every $t > 0$, one has

$$\nu'_{\alpha p'}(t) = \lim_{n\to+\infty} \nu_{\alpha p_n}(t) = \lim_{n\to+\infty} \nu_{p_n}\left(\frac{t}{|\alpha|}\right) = \nu'_{p'}\left(\frac{t}{|\alpha|}\right).$$

Thus (V', ν', τ) is a Šerstnev PN space.

Define $\varphi : V \to V'$ as the function that maps every element $p \in V$ into the constant sequence $\varphi(p) := (p, p, \ldots, p, \ldots)$ thus $\varphi(V) \subset V'$. Moreover,

$$\nu'_{\varphi p} = \lim_{n\to+\infty} \nu_p = \nu_p,$$

which proves that (V, ν, τ) is isometric to a linear subspace of (V', ν', τ).

The image $\varphi(V)$ is a linear subspace dense in (V', ν', τ). Let p' be in V' and $\{p_n\}$ be in p'; then consider the sequence $\{\varphi(p_n)\}$ of elements of V'. Now $\{p_n\}$ is a Cauchy sequence in V, i.e.

$$d_S(\nu_{p_n-p_m}, \varepsilon_0) \to 0 \quad \text{as } n, m \to +\infty$$

so that because of isometry,

$$d_S(\nu_{\varphi(p_n)-\varphi(p_m)}, \varepsilon_0) \to 0 \quad \text{as } n, m \to +\infty.$$

In other words, $\{\varphi(p_n)\}$ is a Cauchy sequence in V'. For every $k \in \mathbf{N}$ one has

$$\nu'_{p'-\varphi(p_k)} = \lim_{n\to+\infty} \nu_{p_n-p_k};$$

and, because of the continuity of d_S, one has, for every $\delta > 0$

$$d_S(\nu'_{p'-\varphi(p_k)}, \varepsilon_0) = \lim_{n\to+\infty} d_S(\nu_{p_n-p_k}, \varepsilon_0) < \delta,$$

if k is large enough.

The space (V', ν', τ) is complete. In fact, let $\{p_n\}$ be a Cauchy sequence of elements from V', i.e.

$$d_S(\nu'_{p'_n-p_m}, \varepsilon_0) \to \qquad \text{as } n, m \to +\infty.$$

Since $\varphi(V)$ is dense in V', for every $n \in \mathbf{N}$ and for every $\eta > 0$, there exists $p_n \in V$ such that

$$d_S(\nu_{\varphi(p_n-p'_n)}, \varepsilon_0) < \eta.$$

Apply the triangle inequality and Lemma 1.3.3 twice in order to obtain

$$\begin{aligned} d_S(\nu_{\varphi(p_n)-\varphi(p_m)}, \varepsilon_0) &\leq d_S(\tau[\nu'_{\varphi(p_n)-p'_n}, \nu'_{p'_n} - \varphi(p_m)], \varepsilon_0) \\ &\leq d_S(\tau^2[\nu'_{\varphi(p_n)-p'_n}, \nu'_{p'_n-p_m}, \nu'_{p'_n} - \varphi(p_m)], \varepsilon_0). \end{aligned}$$

By virtue of the continuity of both τ and d_S it follows that $\{\varphi(p_n)\}$ is a Cauchy sequence in V'; by isometry, $\{(p_n)\}$ is a Cauchy sequence in V. Let p' denote the equivalence class of $\{p_n\}$. Now, for every $n \in \mathbf{N}$,

$$\nu_{p'-\varphi(p_n)} = \lim_{k\to+\infty} \nu_{p_k} - \varphi(p_n),$$

so that, for every $\delta > 0$ and for n large enough,

$$d_S(\nu_{p'-\varphi(p_n)}, \varepsilon_0) = \lim_{k\to+\infty} d_S(\nu_{p_k} - \varphi(p_n), \varepsilon_0) < \delta,$$

which means that $\varphi(p_n) \to p'$. Thus (V', ν', τ) is complete.

In order to show that the completion of (V, ν, τ) is unique up to isometries, let (V', ν', τ) and (V'', ν'', τ) be two different completions of (V, ν, τ) and let $\varphi' : V \to V'$ and $\varphi'' : V \to V'$; be the isometries embedding V into ν' and V'' respectively. Let $p' \in V'$, since $\varphi(V)$ is dense in V', there is a sequence $\{p_n\}$ of elements of V such that $\varphi'(p_n) \to p'$ in V'. By symmetry, $\{p_n\}$ is a Cauchy sequence in V; for the same reason $\{\varphi''(p_n)\}$ is a Cauchy sequence in V''. Since (V'', ν'', τ) is complete, there exists $p'' \in V''$ such that $\varphi''(p_n) \to p''$. Define $\psi : V' \to V''$ via $\psi(p' = p'')$ thus obtaining an isometry of V' and V''. □

Theorem 3.4.2. *Every PN space (V, ν, τ, τ^*) with continuous triangle functions τ and τ^* has a completion, viz. it is isometric to a dense linear subspace of a complete PN space (V, ν, τ, τ^*). This completion is unique up to isometries.*

Proof. The space V' is defined exactly as in the proof of Theorem 3.4.1. One proves in the same way that if p' and q' belong to V', then $p'+q'$ also belongs to V'. Since one can no longer rely on condition (S), a different argument is needed in order to prove that, if α is in $\mathbb{R}$ and p' is in V', then $\alpha p'$ is in V'.

Now assume α is in $\mathbb{R}$ and p' be in V', let $\{p_n\}$ be in p' and consider the sequence $\{\alpha p_n\}$. This latter is a Cauchy sequence in V. This is obviously true for $\alpha = 0$. Because of (N2), one needs to consider only the case $\alpha > 0$.

Arguing by induction, assume first that $\{\alpha p_n\}$ is a Cauchy sequence when α is a positive integer $j < k$. Then

$$\begin{aligned} F_{kp_n,kp_m} &= \nu_{k(p_n-p_m)} \geq \tau(\nu_{p_n-p_m}, \nu_{(k-1)p_n-p_m}) \\ &= \tau(F_{p_n,p_m}, F_{(k-1)p_n,(k-1)p_m}). \end{aligned} \tag{3.4.3}$$

Since τ is continuous and

$$\lim_{n,m\to+\infty} F_{kp_n,kp_m} = \lim_{n,m\to+\infty} F_{(k-1)p_n,(k-1)p_m} = \varepsilon_0,$$

it follows that $\{\alpha p_n\}$ is a Cauchy sequence for every $\alpha \in \mathbf{Z}$. If α is positive, but not an integer, there exists $k \in \mathbf{Z}$ such that $k < \alpha < k+1$. Lemma 3.3.1 now gives

$$F_{(k+1)p_n,(k+1)p_m} \leq F_{\alpha p_n,\alpha p_m} \leq F_{kp_n,kp_m},$$

from which one can immediately conclude that $\{\alpha p_n\}$ is a Cauchy sequence for every $\alpha \geq 0$ and hence for every $\alpha \in \mathbb{R}$. Thus there exists an element $u' \in V'$ to which $\{\alpha p_n\}$ belongs. Define $u' := \alpha p'$. In order to check that this is a good definition, let $\{p_n\} \sim \{\alpha p_n^*\}$. If $\alpha \in [0,1]$, it follows from Lemma 3.3.1 that $F_{p_n,p_n^*} \leq F_{\alpha p_n,\alpha p_n^*}$, since, by assumption F_{p_n,p_n^*} converges to ε_0, and so does $F_{\alpha p_n,\alpha p_n^*}$. If $\alpha = k \in \mathbf{N}$, then Eq. (3.4.3), where p_n^* replaces p_m, holds. Then the same argument yields $\{kp_n\} \sim \{kp_n^*\}$ and, from this it is easily shown that for every $\alpha \in \mathbb{R}$, one has $\{\alpha p_n\} \sim \{\alpha p_n^*\}$.

Therefore V' is a linear space. Only conditions (N2) and (N4) remain to be proved. Proceeding as above, take $p' \in V'$ and let $\{p_n\}$ be a sequence of elements of V that belong to p'; then $\{-p_n\}$ belongs to $-p'$. Since ν' is defined by (3.4.2), one has, on account of property (N2), which holds for ν,

$$\nu'_{-p'} = \lim_{n\to+\infty} \nu_{-p_n} = \lim_{n\to+\infty} \nu_{p_n} = \nu'_{p'}.$$

Moreover, for every $\alpha \in [0,1]$, one has, because τ^* is continuous,

$$\nu'_{p'} = \lim_{n\to+\infty} \nu_{p_n} \leq \lim_{n\to+\infty} \tau^*[\nu_{\alpha p_n}, \nu_{(1-\alpha)p_n}] = \tau^*[\nu'_{\alpha p'}, \nu'_{(1-\alpha)p'}].$$

The remaining part of the proof is similar to that of the corresponding part of the proof of Theorem 3.4.1. □

3.5 Probabilistic Metrization of Generalized Topologies

In Höhle (1977) a problem posed by Thorp about the probabilistic metrization of *generalized topologies* was solved.

We recall some definitions and results that will be used in the next section.

Let V be a non-empty set. A generalized topology (of type T_D) on V is a family of subsets $(\mathcal{U}_p)_{p\in V}$, where $\mathcal{U}_p$ is a filter on V such that $p \in U$ for all $U \in \mathcal{U}_p$. Elements of $\mathcal{U}_p$ are called neighborhoods at p. Such a generalized topology is called *Fréchet-separated* if $\bigcap_{U\in\mathcal{U}_p} U = \{p\}$.

A generalized uniformity $\mathcal{U}$ on V is a filter on $V \times V$ such that every $T \in \mathcal{U}$ contains the diagonal $\{(p,p) : p \in V\}$, and, for all $T \in \mathcal{U}$ one has $T^{-1} := \{(q,p) : (p,q) \in T\}$ also belongs to $\mathcal{U}$. Elements of $\mathcal{U}$ are called *vicinities* (or *entourages*). Every generalized uniformity $\mathcal{U}$ induces a generalized topology as follows: for $p \in V$,

$$\mathcal{U}_p := \{U \subseteq V | \exists T \in \mathcal{U} : U \supseteq \{q \in V | (p,q) \in T\}\}. \tag{3.5.1}$$

A uniformity $\mathcal{U}$ is called *Hausdorff-separated* if the intersection of all vicinities is the diagonal on V.

Theorem 3.5.1. (Höhle, 1977, Theorem 1) *Every Fréchet-separated generalized topology $(\mathcal{U}_p)_{p\in S}$ on a given set S is derivable from a Hausdorff-separated generalized uniformity $\mathcal{U}$ in the sense of* (3.5.1).

Let (V,F) be a PSM space. Consider the system $(\mathcal{N})$ of neighborhoods

$$\mathcal{N} := \cup_{p\in V} \cup_{t>0} \{N_p(t) : p \in V\},$$

called the *strong neighborhood system*, where

$$\begin{aligned} N_p(t) &:= \{q \in V : d_S(F_{pq}, \varepsilon_0) < t\} = \{q \in V : F_{pq}(t) \geq 1 - t\} \\ &= \{q \in V : \nu_{p-q}(t) > 1 - t\} \end{aligned}$$

determines a topology on V, called the *strong topology*. Furthermore, $\mathcal{N}$ admits a countable filter base given by $\{N_p(1/n) : n \in \mathbb{N}\}$, hence the strong neighborhood system is first-countable. All these considerations achieve the following fact as a theorem.

Theorem 3.5.2. *Let (V, F) be a PSM space, then the strong neighborhood system defines a generalized topology of type T_D which is Fréchet-separated and first-countable.*

This generalized topology is called the strong generalized topology of the PSM space (V, F). The main result in Höhle (1977) is the following theorem.

Theorem 3.5.3. *Let T be a t-norm such that $\sup_{0\leq x<1} T(x,x) < 1$. A Fréchet-separated generalized topology $(\mathcal{U}_p)_{p\in V}$ on a set V is derivable from a Menger PM space (V, F, T) if, and only if, there exists a Hausdorff-separated, generalized uniform structure $\mathcal{U}$ having a countable filter base, such that $\mathcal{U}$ is compatible with $(\mathcal{U}_p)_{p\in V}$.*

The following section deals with translation-invariant generalized topologies (TIGT) induced by probabilistic norms.

3.6 TIGT Induced by Probabilistic Norms

Assume that V is a vector space over $\mathbb{R}$. A generalized topology $(\mathcal{U}_p)_{p\in V}$ on V is said to be *translation-invariant* if for all $U \in \mathcal{U}_p$ and $q \in V$, one has $q + U \in \mathcal{U}_{p+q}$. Consequently, a translation-invariant generalized topology is uniquely determined by the neighborhood system $\mathcal{U}_\theta$ at the origin θ of V. In this case, the generalized uniformity from which one can derive the generalized topology is

$$\mathcal{U} := \{T \subseteq V \times V | \exists U \in \mathcal{U}_\theta : T \supseteq \{(p,q) | p - q \in U\}\}.$$

Recall that a subset U of a vector space is called *radial* if $-U = U$; it is called *circled* (or *balanced*) if $\lambda U \subset U$ for all $|\lambda| \leq 1$.

Theorem 3.6.1. *Every PN space (V, ν, τ, τ^*) admits a generalized topology $(\mathcal{U}_p)_{p\in V}$ of type T_D which is Fréchet-separated, translation-invariant, and countable-generated by radial and circled θ-neighborhoods.*

Proof. Let (V, ν, τ, τ^*) be a PN space with τ not necessarily continuous. Let (V, ν) be its associated PSM space, where $F_{p,q} = \nu_{p-q}$. The strong neighborhoods at p are given by

$$N_p(t) = \{q \in V : \nu_{p-q}(t) > 1 - t\} = p + N_\theta(t).$$

In particular, the generalized topology is translation-invariant. By axiom (N1) in Definition 2.2.1, this generalized topology is Fréchet-separated (as in the case of PSM spaces). The countable base of θ-neighborhoods is $\{N_\theta(\frac{1}{n}) : n \in \mathbb{N}\}$, whose elements are clearly radial and circled, by axioms (N2) and (N4) respectively. □

Note that the generalized topology induced by a PN space (V, ν, τ, τ^*) is derivable from the following generalized uniformity:

$$\mathcal{U} := \left\{T \subseteq V \times V | \exists n \in \mathbb{N} : T \supseteq \left\{(p,q) | \nu_{p-q}\left(\frac{1}{n}\right) \geq 1 - \frac{1}{n}\right\}\right\},$$

which is translation-invariant and has a countable filter base of radial and circled vicinities.

Adapting the methods used in Höhle (1977), we show immediately that a converse result holds for such generalized topologies (or generalized uniformities).

Let V be a vector space and $(\mathcal{U}_p)_{p \in V}$ be a Fréchet-separated, translation-invariant, generalized topology of type T_D on V. Then, there is a unique translation-invariant, Hausdorff-separated generalized uniformity, which is defined as follows

$$\mathcal{U} := \{T \subseteq V \times V | \exists U \in \mathcal{U}_\theta : T \supseteq \{(p,q) | p - q \in U\}\}.$$

The analogous result to Theorem 3.5.3 for PN spaces is the following (note that there is an extra assumption on the t-norm T):

Theorem 3.6.2. *Let T be a t-norm such that $\sup_{0 \leq x < 1} T(x,x) < 1$. Suppose $T(x,y) \leq xy$, whenever $x, y \leq \delta$, for some $\delta > 0$. A Fréchet-separated, translation-invariant, generalized topology $(\mathcal{U}_p)_{p \in V}$ on a real vector space V is derivable from a Menger PN space $(V, \nu, \tau_T, \tau_{T^*})$, if, and only if, $\mathcal{U}_\theta$ admits a countable base of radial and circled subsets.*

Proof. The direct implication has been shown above. For the converse, let $\mathcal{B} = \{T_n | n \in \mathbb{N}\}$ be a countable filter base for $\mathcal{U}_\theta$ consisting of radial and circled θ-neighborhoods. Let $n_0 \in \mathbb{N}$ such that $1 - \frac{1}{n_0} \geq \sup_{0 \leq x < 1} T(x,x)$. We can assume that $\frac{1}{n_0} < \delta$, so that $T(x,y) \leq xy$, for all $x, y \leq \frac{1}{n_0}$, where δ is given by hypothesis.

Before defining ν, recall from Höhle (1977, Theorem 2) the d.f.s F_n (used to define the probabilistic metric F):

$$F_n(x) := \begin{cases} 0, & x \le 0, \\ 1 - 1/(n_0(n+1)), & 0 < x \le \frac{1}{n+1}, \\ 1 - 1/(2n_0(n+1)), & \frac{1}{n+1} < x \le 1, \\ 1 - 1/(2^{m+1}n_0(n+1)), & m < x \le m+1 \quad \text{for } m \in \mathbb{N}. \end{cases}$$

By putting $\nu_p(x) = F_{p,\theta}$ we define:

$$\nu_p(x) := \begin{cases} F_0, & p \notin T_1, \\ F_n, & p \in T_n \setminus T_{n+1}, \quad \text{for } n \in \mathbb{N}, \\ \varepsilon_0, & p \in \cap_n T_n. \end{cases}$$

□

We next check that $(V, \nu, \tau_T, \tau_{T^*})$ is a PN space. Axiom (N1) holds because the generalized topology is Fréchet-separated. Axiom (N2) holds because all T_ns are radial. As for (N3) one has

$$\begin{aligned} \tau_T(\nu_p, \nu_q)(x) &= \sup_{r+s=x} \{T(\nu_p(r), \nu_q(s)\} \le 1 - 1/n_0 \le \nu_{p+q}(r+s) \\ &= \nu_{p+q}(x). \end{aligned}$$

As regards (N4): let $p \in T_n$ and $\lambda \in [0,1]$. Then, λp and $(1-\lambda)p$ are also in T_n, because T_n is circled. For $x = r + s$, we have to show that

$$\nu_p(x) \le T^*(\nu_{\lambda p}(r), \nu_{(1-\lambda)p}(s)).$$

Firstly suppose that r and s are strictly greater than 1, $r, s > 1$. Let $a, b, c \in \mathbb{N}$ such that $a < r \le a+1$, $b < s \le b+1$, and $c < r+s \le c+1$. Then,

$$\begin{aligned} \nu_{\lambda p}(r) &= 1 - 1/(2^{a+1}n_0(n+1)), \\ \nu_{(1-\lambda)p}(s) &= 1 - 1/(2^{b+1}n_0(n+1)), \\ \nu_p(r+s) &= 1 - 1/(2^{c+1}n_0(n+1)). \end{aligned}$$

By the properties of T it follows that

$$\begin{aligned} T^*(\nu_{\lambda p}(r), \nu_{(1-\lambda)p}(s)) &= 1 - T(1 - \nu_{\lambda p}(r), 1 - \nu_{(1-\lambda)p}(s)) \\ &= 1 - T(1/(2^{a+1}n_0(n+1)), 1/(2^{b+1}n_0(n+1))) \\ &\ge 1 - (1/(2^{a+1}n_0(n+1))) \cdot (1/(2^{b+1}n_0(n+1))) \\ &\ge 1 - 1/(2^{c+1}n_0(n+1)) \\ &= \nu_p(r+s) = \nu_p(x). \end{aligned}$$

In the third line we have used the fact that the arguments of T are smaller than $1/n_0$, thus we can apply $T(x,y) \le xy$. Then, we obtain $\nu_p \le \tau_{T^*}$

$(\nu_{\lambda p}, \nu_{(1-\lambda)p})$ as desired. The inequality for the other possible values of r and s is checked in a similar way. One concludes that $(V, \nu, \tau_T, \tau_{T^*})$ is a Menger PN space under T.

It only remains to show that the generalized topology induced by ν is the same as the one given at the beginning. We have by construction that

$$T_n = \left\{ p \in V | \nu_p(1/(n+1)) \geq 1 - \frac{1}{n_0(n+1)} \right\}.$$

Thus, the filter base $\{p \in V | \nu_p(\frac{1}{n+1}) \geq 1 - \frac{1}{n+1}\}$ induced by ν is equivalent to $\mathcal{B}$, hence the proof is complete.

Remark 3.6.1. Theorem 3.6.2 also holds if instead of assuming $T(x, y) \leq xy$ near the origin, one assumes that T is Archimedean near the origin (i.e. there is a $\delta > 0$ such that $0 < T(x, x) < x$, for all $0 \leq x \leq \delta$). In that case the d.f. F_n can be chosen as:

$$F_n(x) := \begin{cases} 0, & x \leq 0, \\ 1 - z, & 0 < x \leq \frac{1}{n+1}, \\ 1 - T(z, z), & \frac{1}{n+1} < x \leq 1, \\ 1 - T^{m+1}(z, z) & m < x \leq m+1 \quad \text{for } m \in \mathbb{N}. \end{cases}$$

where $z = 1/(n_0(n+1))$, $T^1(x, y)$ and recursively

$$T^r(x, y) = T(T^{r-1}(x, y), T^{r-1}(x, y)).$$

Chapter 4

Probabilistic Norms and Convergence

4.1 The L^p and Orlicz Norms

The result expressed by Theorem 2.5.2 can be applied to the linear space L^0 of the equivalence classes of random variables $f : \Omega \to \mathbb{R}$. In this case the quotient mapping (2.5.3) is given explicitly by

$$\nu_f(t) := P\{\omega \in \Omega : |f(\omega)| < t\} \quad (t > 0). \tag{4.1.1}$$

As usual in probability theory, we shall write f even when we refer to the equivalence class $\widetilde{f}$ of f.

As a consequence of Theorems 2.5.1 and 2.5.2, (L^0, ν, τ_W) is a Šerstnev PN space, and any linear subspace S of L^0 inherits this property, i.e., $(S, \nu_{|S, \tau_W})$ is a Šerstnev space. Since no confusion is likely to occur we shall keep on denoting by ν the restriction of the probabilistic norm ν to the linear subspace S of V. Important special cases are obtained when S is an L^p space with $p \in [1, +\infty]$ or an Orlicz space. It was proved by Schweizer and Sklar (1980) that all L^p metrics could be derived from a single probabilistic metric. Later this result was extended, with a simplified proof, by Sempi (1985) to the case of Orlicz spaces. However, since both L^p spaces with $p \in [1, +\infty]$ and Orlicz spaces are normed, and are in fact Banach spaces, and since, therefore, their metrics derive from a norm, it is more natural to show that a single probabilistic norm generates the norms of all these spaces.

A simple use of the change of variable formula (see, e.g., Rao, 1987) allows us to prove the following theorem.

Theorem 4.1.1. *Let*

$$L^p(\Omega, \mathcal{A}, P) := \left\{ f \in L^0 : \int |f|^p dP < +\infty \right\} \quad \text{for } p \in [1, +\infty]$$

and

$$L^\infty(\Omega, \mathcal{A}, P) := \left\{ f \in L^0 : \|f\|_\infty := ess\sup |f| < +\infty \right\},$$

then, for all f in L^p or in L^∞, respectively, one has

$$\|f\|_p = \left(\int_{\mathbb{R}_+} t^p \, d\nu_f(t) \right)^{1/p} \tag{4.1.2}$$

and

$$\|f\|_\infty = \sup\{t > 0 : \nu_f(t) < 1\}. \tag{4.1.3}$$

In an Orlicz space L^φ, the Luxemburg norm

$$\|f\|_\varphi := \inf \left\{ k > 0 : \int_\Omega \varphi(f/k) dP \leq 1 \right\},$$

is given by

$$\|f\|_\varphi = \inf \left\{ k > 0 : \int_{\mathbb{R}_+} \varphi\left(\frac{f}{k}\right) d\nu_f \leq 1 \right\}.$$

4.2 Convergence of Random Variables

The study of possible topologies for the various modes of convergence of sequences of random variables has a long history. Let a probability space $(\Omega, \mathcal{A}, P)$ be given and let $L^0(\mathcal{A})$ be the linear space of (equivalence classes) of V-valued random variables, *viz.* measurable functions from Ω to V. Throughout this section $(V, \|\cdot\|)$ will be a normed space.

For a sequence (f_n) of random variables three modes of convergence will be considered: convergence in probability, convergence in L^p with $p \in [1, +\infty]$, and almost sure convergence. The following result is now almost obvious as a consequence of the results on the EN spaces (L^0, ν, τ_W) or (S, ν, τ_W), where S is a linear subspace of L^0 (see Section 2.5). Again, f will denote any function in its equivalence class $\widetilde{f}$.

Theorem 4.2.1. *For a sequence of (equivalence classes of) V-valued random variables $(f_n)_{n\in\mathbb{N}}$, the following statements are equivalent:*

(a) *(f_n) converges in probability to* 0, *viz.* $f_n \xrightarrow[n\to+\infty]{P} 0$;

(b) *the corresponding sequence of probabilistic norms* (ν_{f_n}) *converges weakly to* ε_0, *viz.*

$$d_S(\nu_{f_n}, \varepsilon_0) \xrightarrow[n \to +\infty]{} 0;$$

(c) (f_n) *converges to* 0 *in the strong topology of the Šerstnev space* (L^0, ν, τ_W).

Proof. Since (b) and (c) are equivalent, by definition, it suffices to establish the equivalence of (a) and (b).
The sequence (f_n) converges to the null vector 0 of L^0 if, and only if, for every $t > 0$,

$$\lim_{n \to +\infty} P(|f_n| < t) = 1,$$

or, equivalently, because of the position (4.1.1), if, and only if, for every $t > 0$,

$$\lim_{n \to +\infty} \nu_{f_n}(t) = 1.$$

But this latter statement, in its turn, is equivalent to

$$\lim_{n \to +\infty} d_S(\nu_{f_n}, \varepsilon_0) = 0,$$

which proves the theorem. □

Of course, there is nothing special about 0 as a limit; if one wishes to consider the convergence in probability of the sequence (f_n) to the (V-valued) random variable F, then it is enough to consider the sequence $(f_n - f)$ and its convergence to 0.

Despite the apparent ease with which this result was obtained, it should be noted that it represents an advance on the classical situation. Ky Fan (1944) showed that the topology of convergence in probability can be metrized and introduced the metric on $L^0(\mathcal{A})$ that now bears his name. Later Fréchet (1950) exhibited several metrics that generate this topology. Then Dugué (1955), Marczewski (1955), and Thomasian (1956a; 1956b) proved that, in general, no ordinary norm on L^0 exists that generates the topology of convergence in probability. These results are summarized in Lukacs (1975a). Theorem 4.2.1 shows that this topology is generated by the probabilistic norm (4.1.1). But this same probabilistic norm generates the L^p-topology for $p \in [1, +\infty]$. The following result is an immediate consequence of the representation (4.1.2).

Theorem 4.2.2. *For a sequence of (equivalence classes of) V-valued random variables $(f_n)_{n\in\mathbb{N}}$, and for $p \in [1, +\infty[$, the following statements are equivalent:*

(a) *(f_n) converges to 0 in L^p, viz.* $\xrightarrow[n\to+\infty]{L^p} 0$;

(b) *the sequence of moments of order p of the probabilistic norms (ν_{f_n}) tends to 0, viz.*

$$\int_{\mathbb{R}_+} t^p \, d\nu_{f_n} \xrightarrow[n\to+\infty]{} 0.$$

Now we examine the case of convergence in L^∞.

Theorem 4.2.3. *For a sequence of (equivalence classes of) V-valued random variables $(f_n)_{n\in\mathbb{N}}$, and for $p = +\infty$, the following statements are equivalent:*

(a) *(f_n) converges to 0 in L^∞, viz.* $\xrightarrow[n\to+\infty]{L^\infty} 0$;

(b) *for every $t > 0$, the sequence $(\nu_{f_n}(t))$ is definitely equal to 1, viz. for all $t > 0$, there exists $n_0 = n_0(t) \in \mathbb{N}$ such that $\nu_{f_n}(t) = 1$ if $n \geq n_0$.*

Proof. *(a)* $\Rightarrow$ *(b)*. Assume $\|f_n\|_\infty \to 0$ as n goes to $+\infty$ and let $t > 0$; then, for every $\eta \in]0, t[$ there exists $n_0 = n_0(\eta)$ in $\mathbb{N}$ such that, for every $n \geq n_0$, one has, recalling the representation (4.1.3) of the L^∞ norm,

$$\|f_n'\|_\infty = \sup\{t > 0 : \nu_{f_n}(\eta) < 1\} < \eta,$$

so that $\nu_{f_n}(\eta) = 1$; but then, for all $n \geq n_0$, $\nu_{f_n}(t) \geq \nu_{f_n}(\eta) = 1$.
(b) $\Rightarrow$ *(a)* For $\eta > 0$, let $n_0 = n_0(\eta) \in \mathbb{N}$ be such that $\nu_{f_n}(\eta) = 1$ for all $n \geq n_0$; therefore, because of (4.1.3), $\|f_n\|_\infty < \eta$, namely

$$\lim_{n\to+\infty} \|f_n\|_\infty = 0,$$

which concludes the proof. □

It should be noted that, if the sequence $(\nu_{f_n}(t))$ is definitely equal to 1 for all $t > 0$, then, *a fortiori*, one has

$$\nu_{f_n}(t) \xrightarrow[n\to+\infty]{} 1,$$

or, equivalently,

$$d_S(\nu_{f_n}, \varepsilon_0) \xrightarrow[n\to+\infty]{} 0;$$

this fact is the translation in the language of PN spaces of the fact that convergence in L^∞ implies convergence in probability. The converse is not true, as is well known; in fact, $(\nu_{f_n}(t))$ may well tend to 1 without being definitely equal to 1.

Turning to almost certain convergence, it is to be recalled that, in general, no metric on L^0 exists such that convergence with respect to L^0 coincides with almost certain convergence (Dugué, 1956). Finally Fernique (1988), using and generalizing a famous theorem of Skorohod's (1965), built a probability space and a proper subspace of $L^0(\mathcal{A})$ in which almost sure convergence can be defined through a metric topology. However, even for this convergence, one can construct a suitable PN space in which convergence with respect to the probabilistic norm coincides with almost certain convergence.

Given the probability space $(\Omega, \mathcal{A}, P)$, the set $\mathcal{S} := (L^0(\mathcal{A}))^{\mathbb{N}}$ of all the sequences of (equivalence classes of) V-valued random variables is a real linear space with respect to the component-wise operations: if $s = (f_n)$ and $s' = (g_n)$ are sequences in $\mathcal{S}$ and α is a real number, then the sum $s \oplus s'$ of s and s' and the scalar product $\alpha \odot s$ of α and s are defined respectively via

$$s \oplus s' = (f_n) \oplus (g_n) := (f_n + g_n), \quad \alpha \odot s = \alpha \odot (f_n) := (\alpha f_n).$$

Define $\varphi : \mathcal{S} \to \Delta^+$ by

$$\varphi_s(t) := P\left(\sup_{k\in\mathbb{N}} |f_k| < t\right) = P\left(\bigcap_{k\in\mathbb{N}} |f_k| < t\right) \quad t > 0, s = (f_n). \tag{4.2.1}$$

Theorem 4.2.4. *The triple $(\mathcal{S}, \varphi, \tau_W)$ is a Šerstnev space.*

Proof. Let Θ denote the null sequence $\Theta := (\theta_n)$, where, for every $n \in \mathbb{N}$, $\theta_n = \theta_V$, the null vector of V. Obviously Θ is the null vector of of the linear space $\mathcal{S}$, $\Theta = \theta_{\mathcal{S}}$ and $\varphi_\Theta = \varepsilon_0$.

In the other direction, assume $\varphi_s = \varepsilon_0$, then, one has, for every $t > 0$,

$$\varphi_s(t) := P\left(\sup_{k\in\mathbb{N}} |f_k| < t\right) = 1,$$

so that, for every $t > 0$ and for every $k \in \mathbb{N}$,

$$P(|f_k| < t) = 1;$$

in other words, $\nu_{f_k} = \varepsilon_0$ for every $k \in \mathbb{N}$. But (L^0, ν, τ_W) is a Šerstnev space; therefore, $f_k = \theta_V$ for all $k \in \mathbb{N}$, or, equivalently $s = \Theta$, which concludes the proof of (N1).

Property (N2) is obvious.

Note that for every $\alpha \in \mathbb{R}$, $\alpha \neq 0$, for every $s \in \mathcal{S}$, and for every $t > 0$, one has

$$\varphi_{\alpha \odot s}(t) = P\left(\sup_{k\in\mathbb{N}} |\alpha f_k| < t\right) = P\left(\sup_{k\in\mathbb{N}} |f_k| < \frac{t}{\alpha}\right) = \varphi_s\left(\frac{t}{\alpha}\right).$$

This proves (Š).

For all sequences $s = (f_k)$ and $s' = (g_k)$, for all $t > 0$ and $u \in]0, t[$, one has

$$\begin{aligned} P\left(\bigcup_{k\in\mathbb{N}} (|f_k + g_k| \geq t)\right) &\leq P\left(\bigcup_{k\in\mathbb{N}} (|f_k| + |g_k| \geq t)\right) \\ &\leq P\left[\bigcup_{k\in\mathbb{N}} ((|f_k| \geq u) \cup (|g_k| \geq t - u))\right] \\ &= P\left(\bigcup_{k\in\mathbb{N}} (|f_k| \geq u)\right) + P\left(\bigcup_{k\in\mathbb{N}} (|g_k| \geq t - u)\right) \\ &\quad - P\left[((|f_k| \geq u)) \cap \left(\bigcup_{k\in\mathbb{N}} (|g_k| \geq t - u)\right)\right] \end{aligned}$$

so that

$$\begin{aligned} P\left(\sup_{k\in\mathbb{N}} |f_k + g_k| < t\right) &= P\left(\bigcap_{k\in\mathbb{N}} (|f_k + g_k| < t)\right) \\ &\geq P\left(\bigcap_{k\in\mathbb{N}} (|f_k| + |g_k| < t)\right) \\ &= 1 - P\left(\bigcup_{k\in\mathbb{N}} (|f_k| + |g_k| \geq t)\right) \\ &\geq 1 - P\left(\bigcup_{k\in\mathbb{N}} (|f_k| \geq u)\right) - P\left(\bigcup_{k\in\mathbb{N}} (|g_k| \geq t - u)\right) \\ &\quad + P\left[\left(\bigcup_{k\in\mathbb{N}} (|f_k| \geq u)\right) \cap \left(\bigcup_{k\in\mathbb{N}} (|g_k| \geq t - u)\right)\right] \end{aligned}$$

$$= P\left(\bigcap_{k\in\mathbb{N}}(|f_k| < u)\right) + P\left(\bigcap_{k\in\mathbb{N}}(|g_k| < t-u)\right)$$

$$-1 + P\left[\left(\bigcup_{k\in\mathbb{N}}(|f_k| \geq u)\right) \cap \left(\bigcup_{k\in\mathbb{N}}(|g_k| \geq t-u)\right)\right]$$

$$\geq P\left(\bigcap_{k\in\mathbb{N}}(|f_k| < u)\right) + P\left(\bigcap_{k\in\mathbb{N}}(|g_k| < t-u)\right) - 1.$$

Thus one has, for every $u \in]0, t[$,

$$\varphi_{s\oplus s'}(t) \geq \varphi_s(u) + \varphi_{s'}(t-u) - 1,$$

and hence

$$\varphi_{s\oplus s'} \geq \tau_W(\varphi_s, \varphi_{s'}),$$

which proves (N3) and concludes the proof. □

In order to see how the probabilistic norm (4.2.1) generates almost certain convergence on L^0, let an element s of $\mathcal{S}$ be given, $s = (f_k)_{k\in\mathbb{N}}$ and consider for $n \in \mathbb{N}$ the *n-shift* s_n of s, $s_n := (f_{k+n})_{k\in\mathbb{N}}$, which again belongs to $\mathcal{S}$.

Theorem 4.2.5. *A sequence $s = (f_k)_{k\in\mathbb{N}}$ of V-valued random variables converges almost certainly to θ_V if, and only if, the sequence $(\varphi_{s_n})_{k\in\mathbb{N}}$ of the probabilistic norms of the n-shift of s converges weakly to ε_0, or, equivalently, if, and only if, the sequence (s_n) of the n-shift of s converges to $\theta := (\theta_V, \theta_V, \ldots)$ in the strong topology of $(\mathcal{S}, \varphi, \tau_W)$.*

Proof. All the statements are equivalent to the assertion, which holds for every $t > 0$,

$$1 = \lim_{n\to+\infty} \varphi_{s_n}(t) = \lim_{n\to+\infty} P\left(\bigcap_{k\in\mathbb{N}}(|f_{k+n}| < t)\right)$$

$$= \lim_{n\to+\infty} P\left(\bigcap_{k\geq n}(|f_k| < t)\right) = P\left(\bigcup_{n\in\mathbb{N}}\bigcap_{k\geq n}(|f_k| < t)\right)$$

$$= P\left(\lim_{n\to+\infty} \inf(|f_k| < t)\right).$$

This proves the result. □

Given an element $s = (f_k)$ of $\mathcal{S}$, define $\mu : \mathcal{S} \to \Delta^+$ via

$$\mu_s(t) := \sup_{n\in\mathbb{N}} \varphi_{s_n}(t) = \lim_{n\to+\infty} P\left(\bigcap_{k\geq n}(|f_k| < t)\right);$$

then it follows from the proof of the previous theorem that s converges to θ_V if, and only if, $\mu_s = \varepsilon_0$. An equivalence relationship can be defined in $\mathcal{S}$ by stipulating that, for two sequences $s = (f_k)$ and $s' = (g_k)$, $s \sim s'$ if, and only if, $(f_k - g_k)$ converges almost surely to the null element θ_V.

The strong topology of the Šerstnev space $((L^0(\mathcal{A}))^{\mathbb{N}}, \varphi, \tau_W)$ induces a convergence equivalent to almost sure convergence in $L^0(\mathcal{A})$. This does not contradict the well-known fact that, in general, almost sure convergence does not derive from a topology on $L^0(\mathcal{A})$.

Chapter 5

Products and Quotients of PN Spaces

5.1 Finite Products

The finite product of PN spaces has recently been introduced in Lafuerza-Guillén (2004). In order to define it, one must rely on the concept of *domination*.

Definition 5.1.1. Given a triangle function σ, the *σ-product* of the two PN spaces (V_1, ν_1) and (V_2, ν_2) is the pair $(V_1 \times V_2, \nu^\sigma)$ where $\nu^\sigma : V_1 \times V_2 \to \Delta^+$ is defined, for all $p_1 \in V_1$ and all $p_2 \in V_2$, by

$$\nu^\sigma(p_1, p_2) := \sigma(\nu_1(p_1), \nu_2(p_2)).$$

The following theorem shows how the pair $(V_1 \times V_2, \nu^\sigma)$ can be endowed with a structure of PN space.

Theorem 5.1.1. *Let $(V_1, \nu_1, \tau, \tau^*)$ and $(V_2, \nu_2, \tau, \tau^*)$ be two PN spaces under the same triangle functions τ and τ^*. If σ is a triangle function that dominates τ and is dominated by τ^*, i.e. $\sigma \gg \tau$ and $\tau^* \gg \sigma$, then their σ-product is a PN space under τ and τ^*.*

Proof. Let $\tilde{p} = (p_1, p_2)$ and $\tilde{q} = (q_1, q_2)$ be points in the Cartesian product $V_1 \times V_2$. Then, since σ dominates τ, $\sigma \gg \tau$, one has

$$\begin{aligned}
\nu^\sigma(\tilde{p} + \tilde{q}) &= \sigma(\nu_1(p_1 + q_1), \nu_2(p_2 + q_2)) \\
&\geq \sigma[\tau(\nu_1(p_1), \nu_1(q_1)), \tau(\nu_2(p_2), \nu_2(q_2))] \\
&\geq \tau[\sigma(\nu_1(p_1), \nu_2(p_2)), \sigma(\nu_1(q_1), \nu_2(q_2))] \\
&= \tau(\nu^\sigma(\tilde{p}), \nu^\sigma(\tilde{q})).
\end{aligned}$$

Next, for every $\alpha \in [0,1]$, one has

$$\nu_1(p_1) \le \tau^*(\nu_1(\alpha p_1), \nu_1((1-\alpha)p_1)) \quad \text{and}$$
$$\nu_2(p_2) \le \tau^*(\nu_2(\alpha p_2), \nu_2((1-\alpha)p_2)),$$

so that, since $\tau^* \gg \sigma$,

$$\begin{aligned}\nu^\sigma(\tilde{p}) &= \sigma(\nu_1(p_1), \nu_2(p_2)) \\ &\le [\tau^*(\nu_1(\alpha p_1), \nu_1((1-\alpha)p_1)), \tau^*(\nu_2(\alpha p_2), \nu_2((1-\alpha)p_2))] \\ &\le [\tau^*\sigma(\nu_1(\alpha p_1), \nu_2(\alpha p_2)), \sigma(\nu_1((1-\alpha)p_1), \nu_2((1-\alpha)p_2))] \\ &= \tau^*(\nu^\sigma(\alpha\tilde{p}), \nu^\sigma((1-\alpha)\tilde{p})).\end{aligned}$$

□

Example 5.1.1. The Π_T-product of the PN spaces

$$(V_1, \nu_1, \tau_T, \Pi_M) \quad \text{and} \quad (V_2, \nu_2, \tau_T, \Pi_M)$$

is a PN space under τ_T and Π_M.

If (V_1, ν_1, Π_M) and (V_2, ν_2, Π_M) are equilateral PN spaces with d.f.s F and G respectively, then their Π_M-product is an equilateral space with d.f. given by $\Pi_M(F, G)$ if $F \ne G$, and by F if $F = G$.

If $\tau^* \gg \tau$, both the τ^*-product and the τ-product of $(V_1, \nu_1, \tau, \tau^*)$ and $(V_2, \nu_2, \tau, \tau^*)$ are PN spaces under τ and τ^*. Moreover, if (V_1, ν_1, τ) and (V_2, ν_2, τ) are Šerstnev PN spaces and if $\tau_M \gg \tau$, then both their τ_M-product and their τ-product are Šerstnev PN spaces. Finally, if (V_1, ν_1, T) and (V_2, ν_2, T) are Menger PN spaces under the same continuous t-norm T, then their τ_M-product is a Menger PN space under T.

5.2 Countable Products of PN Spaces

In the case of PM spaces there are two ways of defining countable products of PN spaces (Alsina, 1988). However, there are differences between PN spaces and PM spaces, and, as will be pointed out it is impossible to extend to the case of PN spaces the definitive results of Alsina and Schweizer (1983) on the countable product of PM spaces.

5.2.1 *The Σ-product*

The same definition as in Alsina (1988) can be used in order to define the Σ-product of PN spaces. The treatment of that paper has to be modified in order to take into account that, when dealing with PN spaces, two, rather than only one, triangle functions are to be considered.

Definition 5.2.1. Given a sequence $\left((V_j, \nu_j, \tau_j, \tau_j^*)\right)_{j\in\mathbb{N}}$ of PN spaces and a sequence $(\beta_j)_{j\in\mathbb{N}}$ of strictly positive real numbers such that $\sum_{j\in\mathbb{N}} \beta_j = 1$, their Σ-*product* is the pair (V, ν) where V is the countable Cartesian product of the spaces $\{V_j\}$, $V := \prod_{j\in\mathbb{N}} V_j$, and where, for $p = \{p_j\} \in V$,

$$\nu_p := \sum_{j\in\mathbb{N}} \beta_j \, \nu_j(p_j).$$

Our first result is the following.

Theorem 5.2.1. *Let* $\left((V_j, \nu_j, \tau_j, \tau_j^*)\right)_{j\in\mathbb{N}}$ *be a sequence of PN spaces such that for every* $j \in \mathbb{N}$, $\tau_j \geq \tau_W$ *and* $\tau_j^* \leq \tau_{W*}$. *Then their* Σ-*product* (V, ν) *is a Menger PN space under* W.

Proof. Let $\theta := (\theta_1, \theta_2, \dots, \theta_j, \dots)$, where θ_j is the origin of the vector space V_j $(j \in \mathbb{N})$. Then

$$\nu_\theta = \sum_{j\in\mathbb{N}} \beta_j \, \nu_j(\theta_j) = \sum_{j\in\mathbb{N}} \beta_j \, \varepsilon_0 = \varepsilon_0.$$

In the other direction, if $\nu_p = \varepsilon_0$, then one has $\nu_p(t) = 1$ for every $t > 0$; thus

$$1 = \sum_{j\in\mathbb{N}} \beta_j \, \nu_j(p_j)(t) \leq \sum_{j\in\mathbb{N}} \beta_j = 1,$$

so that one has, for every $j \in \mathbb{N}$ and every $t > 0$, $\nu_j(p_j)(t) = 1$, namely $p_j = \theta_j$ and, therefore, $p = \theta$. This proves (N1). The proof of (N2) is trivial. As for (N3), one has

$$\begin{aligned}
\nu_{p+q} &= \sum_{j\in\mathbb{N}} \beta_j \, \nu_j(p_j + q_j) \geq \sum_{j\in\mathbb{N}} \beta_j \, \tau_j \left(\nu_j(p_j), \nu_j(q_j)\right) \\
&\geq \sum_{j\in\mathbb{N}} \beta_j \, \tau_W \left(\nu_j(p_j), \nu_j(q_j)\right) = \tau_W \left(\nu_p, \nu_q\right).
\end{aligned}$$

Similarly, for every $\alpha \in [0, 1]$ and for every $p \in V$, one has

$$\begin{aligned}
\nu_p &= \sum_{j\in\mathbb{N}} \beta_j \, \nu_j(p_j) \leq \sum_{j\in\mathbb{N}} \beta_j \, \tau_j^* \left(\nu_j(\alpha p_j), \nu_j((1-\alpha)p_j\right) \\
&\leq \sum_{j\in\mathbb{N}} \beta_j \, \tau_{W^*} \left(\nu_j(\alpha p_j), \nu_j((1-\alpha)p_j\right) = \tau_{W^*} \left(\nu_{\alpha p}, \nu_{(1-\alpha)p}\right),
\end{aligned}$$

which establishes (N4) and concludes the proof. □

Corollary 5.2.1. *If $((V_j, \nu_j))_{j\in\mathbb{N}}$ is a sequence of Menger (PN) spaces under W, then their Σ-product is again a Menger PN space under W.*

However, even if $((V_j, \nu_j))_{j\in\mathbb{N}}$ is a sequence of PN spaces with respect to the *same* triangle functions τ and τ^*, their Σ-product need not be a PN space with respect to the same t-norm.

Example 5.2.1 (Alsina, 1978, Example 1.3). Consider the simple space $(\mathbb{R}, |\cdot|, U)$ where U is the d.f. of a random variable uniformly distributed on the interval $(0,1)$ and the sequences p and q in $\prod_{n\in\mathbb{N}} \mathbb{R}$ given by

$$p := \{cn : n \in \mathbb{N}\} \quad \text{and} \quad q := \{n : n \in \mathbb{N}\}.$$

Let $\widetilde{\beta} := \max\{\beta_j : j \in \mathbb{N}\}$ and notice that $\widetilde{\beta} > 0$. Assume, moreover, that the constant $c \in]0, 1/2[$ is small enough to satisfy the inequality

$$\widetilde{\beta} > \frac{2c}{c+1}. \tag{5.2.1}$$

Thus, for every $t > 0$,

$$\nu_{p+q}(t) = \sum_{j\in\mathbb{N}} \beta_j \, U\left(\frac{t}{(c+1)j}\right) < U\left(\frac{t}{c+1}\right).$$

In particular,

$$\nu_{p+q}(1) < \frac{1}{c+1}.$$

On the other hand

$$\nu_p(1/2) = \sum_{j\in\mathbb{N}} \beta_j \, U\left(\frac{1}{2cj}\right) > \widetilde{\beta}\, U\left(\frac{1}{2c}\right) = \widetilde{\beta}$$

since $1/(2c) > 1$, and

$$\nu_q(1/2) = \sum_{j\in\mathbb{N}} \beta_j \, U\left(\frac{1}{2j}\right) > \widetilde{\beta}\, U\left(\frac{1}{c}\right) = \widetilde{\beta},$$

so that

$$\tau_M(\nu_p, \nu_q)(1) \geq \nu_p(1/2) \wedge \nu_q(1/2) \geq \widetilde{\beta}.$$

Therefore one has, because of (5.2.1),

$$\nu_{p+q}(1) < \frac{1}{c+1} < \widetilde{\beta} \leq \tau_M(\nu_p, \nu_q)(1),$$

viz. the Σ-product of the sequence of simple spaces we have considered is not a simple space.

From the point of view of topology, the following result is essential.

Theorem 5.2.2. *The strong topology of the* Σ*-product* $(V, \nu, \tau_W, \tau_{W^*})$ *coincides with the product topology of the sequence of spaces* $\left((V_j, \nu_j, \tau_j, \tau_j^*)\right)_{j\in\mathbb{N}}$.

Proof. The neighborhoods of the type

$$\mathcal{N}_\theta(t) := \{q \in V : \nu_q(t) > 1 - t\}$$

form a basis for the topology of the Σ-product $(V, \nu, \tau_W, \tau_{W^*})$, while the cylinders

$$\prod_{j=1}^{k} \mathcal{N}_{\theta_j}(t_j) \times \prod_{j>k} V_j$$

are a basis for the product topology. Let a neighborhood $\mathcal{N}_\theta(t)$ be given with $t \in]0, 1[$ (if $t \geq 1$, then $\mathcal{N}_\theta(t) = V$). Since there exists $n \in \mathbb{N}$ such that $\sum_{j=1}^n \beta_j > 1 - t$, it is possible to define

$$t' := 1 - \frac{1-t}{\sum_{j=1}^n \beta_j} > 0$$

and let $t_0 := t \wedge t'$. If a point q belongs to the cylinder

$$U_n(t_0) := \prod_{j=1}^{n} \mathcal{N}_{\theta_j}(t_0) \times \prod_{j>n} V_j,$$

then one has $\nu_j(q_j)(t_0) > 1 - t_0$ for every $j = 1, 2, \dots, n$. Now

$$\nu_q(t) \geq \nu_q(t_0) \geq \sum_{j=1}^{n} \beta_j\, \nu_j(q_j)(t_0) > \sum_{j=1}^{n} \beta_j\,(1 - t_0) \geq \sum_{j=1}^{n} \beta_j\,(1 - t') = 1 - t;$$

so that $U_n(t_0) \subset \mathcal{N}_\theta(t)$. In the other direction, consider the cylinder

$$U_n(t_1, t_2, \dots, t_n) := \prod_{j=1}^{n} \mathcal{N}_{\theta_j}(t_j) \times \prod_{j>n} V_j, \tag{5.2.2}$$

and set

$$t := \min\{t_1, t_2, \dots, t_n\} > 0$$

$$t' := 1 - \max\left\{ \beta_j\,(1 - t_j) + \sum_{\substack{k\in\mathbb{N}\\ k\neq j}} \beta_k : j = 1, 2, \dots, n \right\}$$

$$t_0 := t \wedge t'.$$

Then, if q belongs to the neighborhood $\mathcal{N}_\theta(t_0)$, one has, for every $j = 1, 2, \dots, n$,

$$\begin{aligned}\nu_q(t_j) &\geq \nu_q(t_0) > 1 - t_0 \geq 1 - t' \\ &= \max\left\{\beta_j\,(1 - t_j) + \sum_{\substack{k\in\mathbb{N}\\ k\neq j}} \beta_k : j = 1, 2, \dots, n\right\} \\ &\geq \beta_j\,(1 - t_j) + \sum_{\substack{k\in\mathbb{N}\\ k\neq j}} \beta_k \geq \beta_j\,(1 - t_j) + \sum_{\substack{k\in\mathbb{N}\\ k\neq j}} \beta_k\,\nu_k(q_k)(t_j);\end{aligned}$$

this inequality implies

$$\beta_j\,\nu_j(q_j)(t_j) > \beta_j\,(1 - t_j),$$

namely $\nu_j(q_j)(t_j) > 1 - t_j$; therefore

$$\mathcal{N}_\theta(t_0) \subset U_n(t_1, t_2, \dots, t_n);$$

which concludes the proof. □

5.2.2 *The τ-product*

Extending Alsina's construction (1978) a different definition is possible for the countable product of PN spaces. Before proceeding, it is necessary to recall the definition of the $\tau^{(\infty)}$-product of a sequence $(\tau_n)_{n\in\mathbb{N}}$ of triangle functions. One proceeds inductively by setting, for any sequence $(F_n)_{n\in\mathbb{N}}$ of distribution functions in Δ^+, $\tau^{(1)} := \tau_1$, and, for $n \geq 2$,

$$\tau^{(n)}\,(F_1, \dots, F_n, F_{n+1}) := \tau_n\left(\tau^{(n-1)}\,(F_1, \dots, F_n)\,, F_{n+1}\right).$$

The sequence $\left\{\tau^{(n)}\,(F_1, \dots, F_{n+1})\right\}$ is decreasing so that its weak limit always exists, although it may be identically equal to zero; it is denoted by $\tau^{(\infty)}\{F_n\}$,

$$\tau^{(\infty)}\{F_n\} := \text{w-}\lim_{n\to+\infty} \tau^{(n)}\,(F_1, \dots, F_{n+1}).$$

Definition 5.2.2. Given a sequence $\left((V_j, \nu_j, \tau_j, \tau_j^*)\right)_{j\in\mathbb{N}}$ of PN spaces and a continuous triangle function τ, their *τ-product* is the pair (V, ν), where $V = \prod_{j\in\mathbb{N}} V_j$ and $\nu : V \to \Delta^+$ is defined, if $p = (p_n)$ with $p_n \in V_n$ for every $n \in \mathbb{N}$, by

$$\nu_p := \tau^{(\infty)}\left\{\nu_n(p_n)\right\}. \tag{5.2.3}$$

Theorem 5.2.3. *Let $\left((V_j, \nu_j, \tau_j, \tau_j^*)\right)_{j\in\mathbb{N}}$ be a sequence of PN spaces. Assume that there exist two continuous triangle functions τ and τ^* such that*

(a) *for every $n \in \mathbb{N}$, $\tau \leq \tau_n \leq \tau_n^* \leq \tau^*$;*
(b) *τ^* dominates τ, $\tau^* \gg \tau$.*

Then their τ-product (V, ν) is a PN space under τ and τ^*.

Proof. One has $\nu_\theta = \tau^{(\infty)}\{\nu_{\theta_n}\} = \tau^{(\infty)}\{\varepsilon_0, \dots, \varepsilon_0, \dots\} = \varepsilon_0$. Note that this also proves that $\tau^{(\infty)}$ is not identically equal to zero. Assume now $\nu_p = \varepsilon_0$; then one has $\nu_n(p_n) = \varepsilon_0$, and hence $p_n = \theta_n$ for every n; therefore $p = \theta$ and (N1) is proved. Property (N2) is trivial. Let $p = (p_n)$ and $q = (q_n)$ be elements of V. Then, on account of the continuity of τ, one has

$$\begin{aligned}
\nu_{p+q} &= \tau^{(\infty)}\{\nu_n(p_n + q_n)\} \geq \tau^{(\infty)}\{\tau_n(\nu_n(p_n), \nu_n(q_n))\} \\
&\geq \tau^{(\infty)}\{\tau(\nu_n(p_n), \nu_n(q_n))\} \\
&= \tau\left(\tau^{(\infty)}(\nu_n(p_n)), \tau^{(\infty)}(\nu_n(q_n))\right) \\
&= \tau(\nu_p, \nu_q),
\end{aligned}$$

which proves (N3). For every $\alpha \in [0, 1]$, one has, since τ^* is continuous and τ^* dominates τ,

$$\begin{aligned}
\nu_p &= \tau^{(\infty)}\{\nu_n(p_n)\} \leq \tau^{(\infty)}\{\tau_n^*(\nu_n(\alpha p_n), \nu_n((1-\alpha)p_n))\} \\
&\leq \tau^{(\infty)}\{\tau^*(\nu_n(\alpha p_n), \nu_n((1-\alpha)p_n))\} \\
&\leq \tau^*\left(\tau^{(\infty)}\{\nu_n(\alpha p_n)\}, \tau^{(\infty)}\{\nu_n((1-\alpha)p_n)\}\right) \\
&= \tau^*\left(\nu_{\alpha p}, \nu_{(1-\alpha)p}\right),
\end{aligned}$$

which proves (N4) and concludes the proof. □

5.3 Final Considerations

In Alsina and Schweizer (1983) the authors were successful in constructing the countable product of PM spaces in such a way that the following natural, and hence desirable, properties of the product space hold:

(a) if all the PM spaces in the sequence $((S_n, \mathcal{F}_n, \tau))_{n\in\mathbb{N}}$ share the same triangle function τ, then the product space $(S, \mathcal{F})$ is a PM space under the same triangle function τ;
(b) the strong topology of the product space coincides with the product topology.

One naturally wonders whether the construction given in Alsina and Schweizer (1983) may be modified so as to extend it to the case of the countable product of PN, rather than PM, spaces. It will be shown below that the answer to this question must be negative. This is a consequence of the fact that *two* triangle functions, τ and τ^*, rather than only one, appear in the definition of a PN space. Let $((V_j, \nu_j, \tau, \tau^*))_{j\in\mathbb{N}}$ be a sequence of PN spaces under the same two continuous triangle functions τ and τ^*. The same concepts will be used as in Alsina and Schweizer (1983). As in that paper, let $(\beta_j)_{j\in\mathbb{N}}$ be a sequence of strictly positive numbers such that the series $\sum_{j\in\mathbb{N}} \beta_j$ converges. For every $j \in \mathbb{N}$, let φ_j belong to M_{β_j}, the set of strictly increasing functions from $[0, \beta_j]$ onto $[0, +\infty]$. The functions φ_j are requested to be τ-superadditive, *viz.* for all F and G in Δ^+ and for every $j \in \mathbb{N}$,

$$\tau(F, G) \circ \varphi_j \geq \tau\left(F \circ \varphi_j, G \circ \varphi_j\right).$$

Define, for every $j \in \mathbb{N}$, $\widetilde{\nu}(p_j) := \nu(p_j) \circ \varphi_j$ and, for $p = \{p_j : j \in \mathbb{N}\}$ in $\prod_{j\in\mathbb{N}} V_j$,

$$\varphi_p := \tau^\infty\{\widetilde{\nu}(p_j)\}.$$

Then again, with the same proof as in Alsina and Schweizer (1983), one shows that ν_p belongs to $\mathcal{D}^+$ and that the probabilistic norm thus defined satisfies property (N3) of a PN space. The problems arise with property (N4). In fact, in order to satisfy the latter property one has to require that each φ_j be τ^*-*subadditive*, namely, for all F and G in Δ^+ and for every $j \in \mathbb{N}$,

$$\tau^*(F, G) \circ \varphi_j \leq \tau^*\left(F \circ \varphi_j, G \circ \varphi_j\right).$$

This would imply, in the important case of Menger PN spaces under τ_T, where T is a continuous t-norm, that every φ_j is subadditive. But the only functions that are at the same time super- and sub-additive are affine, i.e. of the type $\varphi_j(t) = \alpha_j t + \gamma_j$, and these functions cannot possibly be onto, since $\beta_j < +\infty$ and then $\varphi_j(\beta_j) = \alpha_j \beta_j + \gamma_j < +\infty$. Therefore one may safely state that the problem is still open of defining the countable product of PN spaces in such a way that, if all the spaces are PN space under the same two triangle functions τ and τ^* the product is also a PN space under the same pair of triangle functions, and that the topology of the product space coincides with the product topology.

5.4 Quotients

In the literature devoted to the theory of PN spaces, topological and completeness questions, boundedness and compactness concepts, linear operators, probabilistic norms for linear operators, product spaces, and fixed point theorems have been studied by various authors. However, quotient spaces of PN spaces have never been considered. This section is a first attempt to fill this gap.

We introduce the concept of quotient in PN spaces and prove some theorems with regard to the completeness of a quotient.

According to Definition 12.9.3 in Schweizer and Sklar's book (1983), the following concept exists:

Definition 5.4.1. A triangle function τ is *sup-continuous* if, for every family $\{F_\lambda : \lambda \in \Lambda\}$ of d.f.s in Δ^+ and every $G \in \Delta^+$,

$$\sup_{\lambda\in\Lambda} \tau(F_\lambda, G) = \tau\left(\sup_{\lambda\in\Lambda} F_\lambda, G\right).$$

In view of Lemma 4.3.5 in Schweizer and Sklar's book (1983), this supremum is in Δ^+. An example of a sup-continuous triangle function is τ_T, where T is a left continuous t-norm.

Definition 5.4.2. Let W be a linear subspace of a linear space V and denote by $\sim_W$ a relation on the set V defined via

$$p \sim_W q \Leftrightarrow p - q \in W,$$

for every $p, q \in V$.
Obviously this relationship is an equivalence relation and therefore the set V is partitioned into equivalence classes, $V/\sim_W$.

Proposition 5.4.1. *Let (V, ν, τ, τ^*) be a PN space. Suppose that τ and τ^* are sup-continuous. Let W be a subspace of V and $V/\sim_W$ its quotient defined by means of the equivalence relation $\sim_W$. Let ν' be the restriction of ν to W and define the mapping $\bar{\nu} : V/\sim_W \to \Delta^+$, for all $p \in V$, by*

$$\bar{\nu}_{p+W}(x) := \sup_{q\in W} \{\nu_{p+q}(x)\}.$$

Then, (W, ν', τ, τ^*) is a PN space and $(V/\sim_W, \bar{\nu}, \tau, \tau^*)$ is a PPN space.

Proof. The first statement is immediate. The remainder of the theorem is guaranteed by the fact that W is not necessarily closed in the strong topology.

Notice that by Lemma 4.3.5 in Schweizer and Sklar's book (1983), $\bar{\nu}_{p+W}$ is in Δ^+.

Hereafter we denote by p_W the subset $p+W$ of V, i.e. an element of quotient, and the strong neighborhood of p_W by $N'_{p_W}(t)$. □

Theorem 5.4.1. *Let W be a linear subspace of V. Then the following statements are equivalent:*

(a) *$(V/\sim_W, \bar{\nu}, \tau, \tau^*)$ is a PN space;*
(b) *W is closed in the strong topology of (V, ν, τ, τ^*).*

Proof. Let $(V/\sim_W, \bar{\nu}, \tau, \tau^*)$ be a PN space. For every p in the closure of W and for each $n \in \mathbb{N}$ choose $q_n \in N_p(1/n) \cap W$. Then

$$\bar{\nu}_{p_W}(1/n) = \sup_{q \in W} \nu_{p+q}(1/n) \geq 1 - 1/n,$$

and hence, $d_S(\bar{\nu}_{p_W}, \varepsilon_0) < 1/n$. Thus $p_W = W$ and as a consequence, $p \in W$ and W is closed. Conversely, if W is closed, let $p \in V$ such that $\bar{\nu}_{p_W} = \varepsilon_0$. If $p \notin W$, then $N_p(t) \cap W = \emptyset$, for some $t > 0$. That is to say, for every $q \in W$, $\nu_{p-q}(t) \leq 1 - t$. Therefore $\bar{\nu}_{p_W}(t) = \sup_{q \in W} \nu_{p+q}(t) \leq 1 - 1/n$, which is a contradiction. □

It is of interest to know whether a PN space can be obtained from a PPN space. An affirmative answer is provided by the following proposition.

Proposition 5.4.2. *Let (V, ν, τ, τ^*) be a PPN space and define*

$$C = \{p \in V : \nu_p = \varepsilon_0\}.$$

Then C is the smallest closed subspace of (V, ν, τ, τ^*).

Proof. If $p, q \in C$, then $p + q \in C$ because $\nu_{p+q} \geq \tau(\nu_p, \nu_q) = \varepsilon_0$. Now suppose $p \in C$. For $\alpha \in [0,1]$ one has $\nu_{\alpha p} \geq \nu_p$ by Lemma 3.3.1. For $\alpha > 1$, let $k = [\alpha] + 1$. Then, using the iterates of (N3) in Definition 2.2.1 one has $\nu_{kp} \geq \tau^{k-1}(\nu_p, \cdots, \nu_p) = \varepsilon_0$. By Lemma 3.3.1 $\nu_{\alpha p} \geq \nu_{kp}$. As a consequence, αp belongs to C for all $\alpha \in \mathbb{R}$. Furthermore it is easy to check that the set C is closed because of the continuity of the probabilistic norm, ν. Now, let W be a closed linear subspace of V and $p \in C$. Suppose that for some $t > 0$, $N_p(t) \cap W = \emptyset$, then $\nu_p(t) \leq 1 - t$, which is a contradiction, hence $C \subseteq W$. □

Remark 5.4.1. Moreover, with V and C as in Proposition 5.4.2, for all $p \in V$ and $r \in C$, one has

$$\bar{\nu}_{pW} \geq \nu_p = \nu_{p+r-r} \geq \tau(\nu_{p+r}, \nu_{-r}) = \nu_{p+r}.$$

Thus the probabilistic norm $\bar{\nu}$ in $(V/\sim_C, \bar{\nu}, \tau, \tau^*)$ coincides with that of (V, ν, τ, τ^*).

Example 5.4.1. Let (V, ν, T) be a Menger PN space. Suppose that W is a closed subspace of V, and $V/\sim_W$ its quotient. Then (W, ν', T) and $(V/\sim_W, \bar{\nu}, T)$ are Menger PN spaces.

Theorem 5.4.2. *Let (V, ν, τ, τ^*) be a PN space. Suppose that τ and τ^* are sup-continuous. Let W be a closed subspace of V with respect to the strong topology of (V, ν, τ, τ^*). Let*

$$\pi : V \to V/\sim_W$$

be the canonical projection. Then π is strongly bounded, open, and continuous with respect to the strong topology of (V, ν, τ, τ^*) and $(V/\sim_W, \nu, \tau, \tau^*)$. In addition, the strong topology and the quotient topology on $V/\sim_W$, induced by π, coincide.

Proof. One has that $\bar{\nu}_{pW} \geq \nu_p$ which implies π is strongly bounded and, by Theorem 8.1.3, also continuous.
The map π is open because of the equality $\pi(N_p(t)) = N'_{pW}(t)$. □

Example 5.4.2. Let $(V, \|\cdot\|)$ be a normed space and define $\nu : V \to \Delta^+$ via $\nu_p := \varepsilon_{\|p\|}$ for every $p \in V$. Let τ, τ^* be continuous triangle functions such that $\tau \leq \tau^*$ and $\tau(\varepsilon_a, \varepsilon_b) = \varepsilon_{a+b}$, for all $a, b > 0$. For instance, it suffices to take $\tau = \tau_T$ and $\tau^* = \tau_{T^*}$, where T is a continuous t-norm and T^* is its t-conorm. Then (V, ν, τ, τ^*) is a Menger PN space.

Assume that τ is sup-continuous. Let W be a closed linear subspace of V with respect to the strong topology of (V, ν, τ, τ^*). By Theorem 5.4.1, $(V/\sim_W, \bar{\nu}, \tau, \tau^*)$ is a PN space in which $\bar{\nu}_{pW} = \sup_{w \in W} \varepsilon_{\|p+w\|}$. On the other hand, if one considers the normed space $(V/\sim_W, \|\cdot\|')$, where $\|pW\|' = \inf_{w \in W} \|p + w\|$, then one can easily prove that the PN structure given to the normed space $(V/\sim_W, \|\cdot\|')$ by means of $\bar{\nu}_{pW} := \varepsilon_{\|pW\|'}$ coincides with $(V/\sim_W, \bar{\nu}, \tau, \tau^*)$.

5.4.1 *Completeness results*

The completeness results in a quotient PN space have recently been introduced in Lafuerza-Guillén, O'Regan, and Saadati (2007). When a PN space (V, ν, τ, τ^*) is strongly complete, then we say that it is a probabilistic normed Banach (PNB) space.

Lemma 5.4.1. *Given the PN space $(V/\sim_W, \bar{\nu}, \tau, \tau^*)$ in which τ and τ^* are sup-continuous, let W be a closed subspace of V. The following statements hold:*

(i) *If $p \in V$, then for every $\varepsilon > 0$ there is a p' in V such that $p'+W = p+W$ and*

$$d_S(\nu_{p'}, \varepsilon_0) < d_S(\bar{\nu}_{p+W}, \varepsilon_0) + \varepsilon;$$

(ii) *If p is in V and $\bar{\nu}_{p+W} \geq G$ for some d.f. $G \neq \varepsilon_0$, then there exists $p' \in V$ such that $p + W = p' + W$ and $\nu_{p'} \geq \tau(\bar{\nu}_{p+W}, G)$.*

Proof. (i) We know

$$\bar{\nu}_{p+W} = \sup\{\nu_{p-q} : q \in W\}.$$

Now, let q be an element of W such that

$$\bar{\nu}_{p+W} < \nu_{p-q} + \frac{\varepsilon}{2}.$$

We put $p - q = p'$. One has

$$\begin{aligned} d_S(\bar{\nu}_{p+W}, \varepsilon_0) &= \inf\{h > 0 : d_S(\bar{\nu}_{p+W}(h^+) > 1 - h\} \\ &\geq \inf\left\{h > 0 : \nu_{p'}(h^+) + \frac{\varepsilon}{2} > 1 - h\right\} \\ &= \inf\left\{h > 0 : \nu_{p'}(h^+) > 1 - \left(h + \frac{\varepsilon}{2}\right)\right\} \\ &\geq \inf\left\{h > 0 : \nu_{p'}\left(\left(\left(h + \frac{\varepsilon}{2}\right)^+\right) > 1 - \left(h + \frac{\varepsilon}{2}\right)\right\} \\ &> d_S(\nu_{p'}, \varepsilon_0) - \varepsilon. \end{aligned}$$

(ii) Because of the definition of supremum and sup-continuity of τ, there exists a $q_n \in W$ such that $q_n \to q$ if $n \to +\infty$ and

$$\nu_{p+q_n} > \tau(\bar{\nu}_{pW}, \varepsilon_0) - \frac{1}{n} \geq \tau(\bar{\nu}_{pW}, G) - \frac{1}{n}.$$

Now it is enough to put $p' = p + q$ and see that, when $n \to +\infty$, one has $\nu_{p+q} \geq \tau(\bar{\nu}_{pW}, G)$.

Let p, q be elements of V such that $d_S(\nu_{(p-q)+W}, \varepsilon_0) < \delta$ for some positive δ. There is a $q' \in V$ such that $(p - q') + W = (p - q) + W$ and

$$d_S(\nu_{p-q'}, \varepsilon_0) < \delta.$$

□

Theorem 5.4.3. *Let W be a closed subspace of V and suppose that (V, ν, τ, τ^*) is a PNB space with τ and τ^* sup-continuous. Then, $(V/\sim_W, \bar{\nu}, \tau, \tau^*)$ is also a PNB space.*

Proof. Let (a_n) be a strong Cauchy sequence in $(V/\sim_W, \bar{\nu}, \tau, \tau^*)$, i.e. for every $\delta > 0$, there exists $n_0 = n_0(\delta) \in \mathbb{N}$ such that, for all $m, n > n_0$, $d_S(\bar{\nu}_{a_n - a_m}, \varepsilon_0) < \delta$.

Now, define a strictly decreasing sequence (δ_n) with $\delta_n > 0$ in the following way: let $\delta_1 > 0$ be such that

$$\tau(B_{d_S}(\varepsilon_0; \delta_1) \times B_{d_S}(\varepsilon_0; \delta_1)) \subseteq B_{d_S}(\varepsilon_0; 1),$$

where $B_{d_S}(\varepsilon_0; \lambda) = \{F \in \Delta^+; d_S(F, \varepsilon_0) < \lambda\}$. For $n > 1$, define δ_n by induction in such a manner that

$$\tau(B_{d_S}(\varepsilon_0; \delta_n) \times B_{d_S}(\varepsilon_0; \delta_n)) \subseteq B_{d_S}\left(\varepsilon_0; \min\left(\frac{1}{n}, \delta_{n-1}\right)\right). \tag{5.4.1}$$

There is a subsequence (a_{n_i}) of (a_n) with

$$d_S(\bar{\nu}_{a_{n_i} - a_{n_{i+1}}}, \varepsilon_0) < \delta_{i+1}. \tag{5.4.2}$$

Because of the definition of the canonical projection π one can say that $\pi^{-1}(N'_{p_W}(t)) = N_p(t)$ and consequently $\pi^{-1}(a_{n_i}) = x_i$ exists. Inductively, from Lemma 5.4.1 we can find $x_i \in V$ such that $\pi(x_i) = a_{n_i}$ and then

$$d_S(\nu_{x_i - x_{i+1}}, \varepsilon_0) < \delta_{i+1} \tag{5.4.3}$$

holds. We claim that (x_i) is a strong Cauchy sequence in (V, ν, τ, τ^*). By applying the relations (5.4.1), (5.4.2), and (5.4.3) to $i = m-1$ and $i = m-2$, and using Lemma 1.3.3, one obtains the inequalities

$$\begin{aligned} d_S(\nu_{x_m - x_{m-2}}, \varepsilon_0) &\leq d_S(\tau(\nu_{x_{m-1} - x_m}, \nu_{x_{m-2} - x_{m-1}}), \varepsilon_0) \\ &< \min\left(\frac{1}{m-1}, \delta_{m-2}\right). \end{aligned}$$

Following this reasoning, we obtain that $d_S(\nu_{x_m - x_n}, \varepsilon_0) < 1/n$ and therefore, x_i is a strong Cauchy sequence. Since it was assumed that (V, ν, τ, τ^*) is strongly complete, the sequence (x_i) is strongly convergent and hence, by the continuity of π, (a_{n_i}) is also strongly convergent. From this and taking into account the continuity of τ and Lemma 1.3.3, one sees that the whole sequence (a_n) strongly converges. □

The converse of the above theorem also holds.

Theorem 5.4.4. *Let (V, ν, τ, τ^*) be a PN space in which τ and τ^* are sup-continuous, and let $(V/\sim_W, \bar{\nu}, \tau, \tau^*)$ be its quotient space with respect to the closed subspace W. If any two of the three spaces V, W, and $V/\sim_W$ are strongly complete, so is the third.*

Proof. If V is a strongly complete PN space, so are W and $V/\sim_W$. Therefore all one needs to check is that V is strongly complete whenever both W and $V/\sim_W$ are strongly complete. Suppose W and $V/\sim_W$ are strongly complete PN spaces and (p_n) be a strong Cauchy sequence in V. Since

$$\bar{\nu}_{(p_m - p_n)} + W \geq \nu_{p_m - p_n}$$

whenever $m, n \in \mathbb{N}$, the sequence $(p_n + W)$ is strong Cauchy in $V/\sim_W$ and therefore, it converges strongly to $q + W$ for some $q \in V$. Thus there exists a sequence of d.f.s (H_n) such that $H_n \to \varepsilon_0$ and $\bar{\nu}_{(p_n - q)+W} > H_n$. Now by Lemma 5.4.1 there exists (q_n) in V such that $q_n + W = (p_n - q) + W$ and

$$\nu_{q_n} > \tau(\bar{\nu}_{(p_n - q)+W}, H_n).$$

Thus $\nu_{q_n} \to \varepsilon_0$. Therefore $(p_n - q_n - q)$ is a strong Cauchy sequence in W and is strongly convergent to a point $r \in W$ and implies that (p_n) converges strongly to $r + q$ in V. Hence V is strongly complete. □

Theorem 5.4.5. *Let $(V_1, \nu^1, \tau, \tau^*), \cdots, (V_n, \nu^n, \tau, \tau^*)$ be PNB spaces in which τ and τ^* are sup-continuous. Suppose that there is a triangle function σ such that $\tau^* \gg \sigma$ and $\sigma \gg \tau$. Then their σ-product is a PNB space.*

Proof. One proves for $n = 2$ (see Theorem 5.1.1) and then we apply induction for an arbitrary n. Since the quotient norm of

$$\frac{V_1 \times V_2}{V_1 \times \theta_2}(\simeq V_2)$$

is the same as ν^2 and the restriction of the product norm of $V_1 \times V_2$ to $V_1 \times \theta_2(\simeq V_1)$ is the same as ν^1, in view of Theorem 5.4.3, the proof is complete. □

By Theorem 5.4.2 one has the following corollaries.

Corollary 5.4.1. *Under the assumptions of Proposition 5.4.1 and that W is a closed subset of V, the probabilistic norm $\bar{\nu} : V/\sim_W \to \Delta^+$ in $(V/\sim_W, \bar{\nu}, \tau, \tau^*)$ is uniformly continuous.*

Proof. Let η be a positive real number, $\eta > 0$. By Theorem 5.4.2 there exists a pair (p', q') in $(V \times V)$ such that $d_S(\bar{\nu}_{\pi(p-p')}, \varepsilon_0) < \eta$ and $d_S(\bar{\nu}_{\pi(q-q')}, \varepsilon_0) < \eta$, whenever $d_S(\nu_{p-p'}, \varepsilon_0) < \eta$ and $d_S(\nu_{q-q'}, \varepsilon_0) < \eta$. On the other hand, we have

$$\bar{\nu}_{\pi(p-p')} \geq \tau(\tau(\bar{\nu}_{\pi(p-p')}, \bar{\nu}_{\pi(q-q')}), \bar{\nu}_{\pi(p-q)})$$

and

$$\bar{\nu}_{\pi(p-q)} \geq \tau(\tau(\bar{\nu}_{\pi(p-p')}, \bar{\nu}_{\pi(q-q')}), \bar{\nu}_{\pi(p'-q')}).$$

Thus from the relationship (12.1.5) in Schweizer and Sklar (1983) and Lemma 3.2.1 it follows that for any $h > 0$ there is an appropriate $t > 0$ such that

$$d_S(\bar{\nu}_{\pi(p-q)}, \bar{\nu}_{\pi(p'-q')}) < h,$$

whenever $p' \in N_p(\eta)$ and $q' \in N_q(\eta)$. This implies that $\bar{\nu}$ is a uniformly continuous mapping from $V/\sim_W$ into Δ^+. Also the inequality

$$d_S(\bar{\nu}_{\pi((p+q)-(p'+q'))}, \varepsilon_0) \leq d_S(\nu_{(p+q)-(p'+q')}, \varepsilon_0)$$

implies that $(V/\sim_W, +)$ is a topological group. □

Corollary 5.4.2. *Let (V, ν, τ, τ^*) be a PN space such that τ^* is Archimedean, τ and τ^* are sup-continuous, and $\nu_p \neq \varepsilon_\infty$ for all $p \in V$. If we define a quotient probabilistic norm via Proposition 5.4.1, then $(V/\sim_W, \bar{\nu}, \tau, \tau^*$ is a PPN space where the scalar multiplication is a continuous mapping from $\mathbb{R} \times V/\sim_W$ into $V/\sim_W$.*

Chapter 6

$\mathcal{D}$-Boundedness and $\mathcal{D}$-Compactness

6.1 The Probabilistic Radius

In order to introduce the various notions of boundedness the definition of probabilistic radius will be needed.

Definition 6.1.1. The *probabilistic radius* R_A of a non-empty set A in PN space (V, ν, τ, τ^*) is defined by

$$R_A := \begin{cases} \ell^- \varphi_A(t), & t \in [0, +\infty[\\ 1, & t = +\infty \end{cases}$$

where $\varphi_A(t) := \inf\{\nu_p(t) : p \in A\}$.

The two results of the next theorem present properties of the probabilistic radius. The first of them is just the analogue of a classical result, while the second one generalizes the well-known relationship $r_{A \cup B} \leq r_A + r_B$ of the set A and B.

Theorem 6.1.1. *In a PN space (V, ν, τ, τ^*) the probabilistic radius has the following properties:*

(a) *the probabilistic radius R_A of every non-empty set A is equal to the probabilistic radius of its closure $\overline{A}$, $R_A = R_{\overline{A}}$;*

(b) *if the sets A and B are non-empty, then*

$$R_{A \cup B} \geq \tau(R_A, R_B) \tag{6.1.1}$$

and

$$R_{A+B} \geq \tau(R_A, R_B) \tag{6.1.2}$$

where $A + B := \{p + q : p \in A, q \in B\}$.

Proof.

(a) Because one has $A \subset \bar{A}$ and as a consequence, $R_A \geq R_{\bar{A}}$, only the converse inequality $R_A \leq R_{\bar{A}}$ has to be proved. We proved in Theorem 3.2.1 that the probabilistic norm $\nu : V \to \Delta^+$ is uniformly continuous when the PN space (V, ν, τ, τ^*) is endowed with the strong topology of the metric d_S. This means that, for every $\eta > 0$, there exists $\delta = \delta(\eta) > 0$ such that $d_S(\nu_p, \nu_q) < \eta$ whenever $d_S(\nu_{p-q}, \varepsilon_0) < \delta$. Now, for every $p \in \bar{A}$, there exists $q(p) \in A$ such that

$$d_S(\nu_{p-q(p)}, \varepsilon_0) < \delta;$$

therefore $d_S(\nu_p, \nu_q) < \eta$. In particular, for every $t \in]0, 1/\eta[$, one has

$$\nu_p(t) \geq \nu_{q(p)}(t - \eta) - \eta.$$

Then, for $t \in]0, 1/\eta[$, one has

$$\begin{aligned}\varphi_{\bar{A}}(t) &= \inf_{p \in \bar{A}} \nu_p(t) \geq \inf_{p \in \bar{A}} \nu_{q(p)}(t - \eta) - \eta \\ &= \inf_{q \in A} \nu_q(t - \eta) - \eta \geq \inf_{q \in A} \nu_q(t - \eta) - \eta = \varphi_A(t - \eta) - \eta.\end{aligned}$$

(b) For every $p \in A \cup B$ and for every $q \in B$, we have that

$$\nu_p = \tau(\nu_p, \varepsilon_0) \geq \tau(\nu_p, \nu_q) \geq \tau(\nu_p, R_B),$$

because $R_B \leq \nu_q$ for all $q \in B$. Therefore, if p is in A, we have $\nu_p \geq (R_A, R_B)$.
The same argument for p in $A \cup B$ and q in A leads to the inequality $\nu_p \geq \tau(R_A, R_B)$.
The assertion now follows from the last two inequalities.
Similarly, if $p \in A$ and $q \in B$, one has

$$\nu_{p+q} \geq \tau(\nu_p, \nu_q) \geq \tau(R_A, R_B),$$

whence $R_{A+B} \geq \tau(R_A, R_B)$.
Alongside the probabilistic radius also the truncated radius of a set A may be introduced. □

Definition 6.1.2. Given $\alpha \in [0, +\infty[$, the α-level set of the probabilistic norm $\nu_p (p \in V)$ is defined by

$$\nu_p^\alpha(t) := \begin{cases} 0, & t \leq \alpha \\ \nu_p(t), & t > \alpha. \end{cases} \tag{6.1.3}$$

Then the *α-truncated radius* R_A^α is defined by

$$R_A^\alpha := \begin{cases} 0, & t \le \alpha \\ l^- \inf\{\nu_p^\alpha(t) : p \in A\}, & t > \alpha. \end{cases} \tag{6.1.4}$$

It is immediately checked that

$$R_A^\alpha(t) := \begin{cases} 0, & t \le \alpha \\ l^- \inf\{\nu_p^\alpha(t) : p \in A\}, & t > \alpha. \end{cases}$$

6.2 Boundedness in PN Spaces

Different kinds of bounded sets may be introduced in a PN space; this situation contrasts sharply with what happens in a 'classical' normed space $(V, \|\cdot\|)$ where a set A is either bounded if there exists a constant $k > 0$ such that $\|p\| \le k$ for all p in A, or unbounded. The results of this section, as well as most of the next one, are in papers by Lafuerza-Guillén *et al.* (1999, 1998), Lafuerza-Guillén (2001), Lafuerza-Guillén *et al.* (2010) and Lafuerza-Guillén *et al.* (2012).

Definition 6.2.1. A non-empty set A in PN space (V, ν, τ, τ^*) is said to be

(a) *certainly bounded*, if $R_A(t_0) = 1$ for some $t_0 \in]0, +\infty[$;
(b) *perhaps bounded*, if one has $R_A(t) < 1$ for every $t \in]0, +\infty[$, but

$$\lim_{t \to +\infty} R_A(t) = 1;$$

(c) *perhaps unbounded*, if $R_A(t_0) > 0$ for some $t_0 \in]0, +\infty[$ and

$$\lim_{t \to +\infty} R_A(t) \in]0, 1[;$$

(d) *certainly unbounded*, if $\lim_{t \to +\infty} R_A(t) = 0$, i.e., if $R_A = \varepsilon_\infty$. Moreover, the set A will be said to be *distributionally bounded* (henceforth $\mathcal{D}$-*bounded*) if either (a) or (b) holds, i.e., if $R_A \in \mathcal{D}^+$; otherwise, i.e., if R_A belongs to $\Delta^+ \setminus \mathcal{D}^+$, A will be said to be $\mathcal{D}$-*unbounded*.

The following lemma is a simple rephrasing of the definition in terms of φ_A instead of the probabilistic radius R_A.

Lemma 6.2.1. *Let A be a non-empty set in a PN space (V, ν, τ, τ^*). Then*

(a) *A is certainly bounded if, and only if, $\varphi_A(t_0) = 1$ for some $t_0 \in]0, +\infty[$;*
(b) *A is perhaps bounded if, and only if, $\varphi_A(t) < 1$ for every $t \in]0, +\infty[$ and $\ell^- \varphi_A(+\infty) = 1$;*

(c) *A is perhaps unbounded if, and only if,* $\ell^- \varphi_A(+\infty) \in]0,1[$;
(d) *A is certainly unbounded if, and only if,* $\ell^- \varphi_A(+\infty) = 0$, *i.e.*, $\varphi_A(+\infty) = \varepsilon_\infty$.

As a justification of the preceding definition, the reader may think of the value at t of the probabilistic norm ν_p of p as the probability that the norm $\|p\|$ is smaller than t. Then a set A is certainly bounded if, and only if, there exists $t_0 > 0$ such that, with probability 1, $\|p\| < t_0$; thus, A is almost certainly included in the open ball $B(t_0)$ centred at the origin θ and of radius t_0. This closely corresponds to the idea of what a bounded set is in probabilistic terms. If the set A is not certainly bounded, then it is perhaps bounded if, and only if, for every $\delta > 0$, there exists $t_0 = t_0(\delta) > 0$ such that every point p in A belongs to $B(t_0)$ with probability greater than $1 - \delta$. The set A is certainly unbounded if, and only if, for every $\delta > 0$ and for every $t_0 > 0$, there exists a point p in A that lies outside $B(t_0)$ with probability greater than $1 - \delta$. Finally, if A is not certainly unbounded, then it is perhaps unbounded if, and only if, there exists $\delta \in [0,1]$ such that, for every $t_0 > 0$, there is a point p in A that lies outside $B(t_0)$ with probability greater than δ.

The proof of the following result is immediate.

Theorem 6.2.1. *A non-empty set A in a PN space (V, ν, τ, τ^*) is* distributionally bounded (*henceforth $\mathcal{D}$-bounded*) *if, and only if, there exists a proper d.f. $G \in \mathcal{D}^+$ such that $\nu_p \geq G$ for every $p \in A$.*

It follows at once from Definition 6.2.1 *that a set A is $\mathcal{D}$-bounded if, and only if,*

$$\lim_{t \to +\infty} \varphi_A(t) = 1. \tag{6.2.1}$$

This latter condition implies that

$$\lim_{t \to +\infty} \nu_p(t) = 1 \quad \text{for every } p \in A. \tag{6.2.2}$$

The converse is not true as the following example shows.

Example 6.2.1. Let $(\mathbb{R}, |\cdot|)$ be the set of the real numbers endowed with the usual norm. It can be made into a Menger space as in Example 2.2.2 by choosing any continuous t-norm T. Let A be any unbounded set of $\mathbb{R}$. Then for every $t \in]0, +\infty[$, there exists $p \in A$ such that $|p| \geq t$. Consequently

$$\varphi_A(t) = \inf_{p \in A} \nu_p(t) = \varepsilon_{|p|}(t) = 0,$$

whence $\lim_{t \to +\infty} \varphi_A(t) = 0$; thus (6.2.1) is not satisfied. On the other hand, one has, for every $p \in A$,

$$\lim_{t \to +\infty} \nu_p(t) = \lim_{t \to +\infty} \varepsilon_{|p|}(t) = 1,$$

so that (6.2.2) is satisfied.

The following result provides a characterization of some of the boundedness concepts introduced in Definition 6.2.1 in terms of the truncated radius.

Theorem 6.2.2. *A nonempty set A of a PN space (V, ν, τ, τ^*) is*

(a) *certainly bounded if, and only if, there exists $\alpha \in]0, +\infty[$ such that*

$$R_A^{(\alpha)} = \varepsilon_\alpha;$$

(b) *perhaps bounded if, and only if, one has, for every $\alpha \in]0, +\infty[$,*

$$R_A^{(\alpha)} < \varepsilon_\alpha \quad \text{and} \quad \ell^- R_A(+\infty) = 1.$$

Proof.

(a) Let A be certainly bounded. Then, there exists x_0 such that $R_A(x_0) = 1$; as a consequence, $R_A^{x_0} = \varepsilon_0$. In the other direction, assume $R_A^\alpha = \varepsilon_\alpha$; then

$$R_A(\alpha+) = R_A^\alpha(\alpha+) = \varphi_A(\alpha) = 1,$$

so that A is certainly bounded.

(b) One has $R_A(t) < 1$ for every $t \in]0, +\infty[$, if, and only if, $R_A^{(\alpha)} < \varepsilon_\alpha$.

□

We now examine the boundedness concepts in the special PN spaces of Sections 2.3, 2.4, and 2.5.

Example 6.2.2. Let (V, G, Π_M) be an equilateral PN space. If there is a point t_0 in $]0, +\infty[$ such that $G(t_0) = 1$, then every non-empty set of V is certainly bounded; otherwise, only the singleton $\{\theta\}$ is certainly bounded. For every subset A of V, one has $\varphi_A = G$ so that the set A is perhaps bounded if, and only if, $\ell^- G(+\infty) = 1$; if $\ell^- G(+\infty) < 1$, then A is perhaps unbounded.

Example 6.2.3. Let $(V, \|\cdot\|)$ be a normed space and consider the distance d.f. G and the simple Menger PN space $(V, \|\cdot\|, G, M)$. Then

(a) If there exists a point x_0 in $]0, +\infty[$ such that $G(x_0) = 1$, then the certainly bounded sets of $(V, \|\cdot\|, G, M)$ coincide with bounded sets of the normed space $(V, \|\cdot\|)$. In fact, if $\|p\| \leq k$ for every $p \in A$, the probabilistic radius of A is

$$R_A(t) = \ell^- \inf\left\{G\left(\frac{t}{\|p\|}\right)\right\} = G\left(\frac{t}{k}\right);$$

Thus it suffices to take $t_0 = kx_0$ in order to have $R_A(t_0) = 1$. In the other direction, if $R_A(t_0) = 1$, it means that, for every p in A, one has

$$G\left(\frac{t}{\|p\|}\right) = \nu_p(t_0) = 1;$$

therefore $\|p\|$ cannot tend to $+\infty$ and there must exist a constant $k > 0$ such that $\|p\| \leq k$. Moreover an unbounded set A in $(V, \|\cdot\|)$ is either perhaps unbounded or certainly unbounded in the PN space $(V, |\cdot\|, G, M)$ according to whether

$$\ell^+ G(0) := \lim_{t \to 0^+} G(t)$$

belongs to the open interval $]0, 1[$ or is equal to 0, respectively.

(b) If $\ell^- G(+\infty) = 1$, but $G(t) < 1$ for every $t \in]0, +\infty[$, then the singleton θ is the only certainly bounded set of $(V, \|\cdot\|, G, M)$; the perhaps bounded sets of $(V, \|\cdot\|, G, M)$ coincide with the bounded sets of $(V, \|\cdot\|)$, while the unbounded sets of $(V, \|\cdot\|)$, are either perhaps unbounded or certainly unbounded in the PN space $(V, \|\cdot\|, G, M)$ according to whether $\ell^+ G(0) > 0$ or $\ell^+ G(0) = 0$, respectively.

(c) If $\ell^- G(+\infty)$ is in $]0, 1[$, everything behaves as in the previous case, the only difference being that the bounded sets of $(V, \|\cdot\|)$ different from θ are perhaps unbounded in $(V, \|\cdot\|, G, M)$.

The same result holds for the α-simple spaces.

Example 6.2.4. The description of the various types of boundedness of Definition 6.2.1 is particularly transparent in the case of EN spaces (see Section 2.5); and here the motivation behind the definition also comes to the surface. Let A be a subset of an EN space (S, ν), i.e. a subset of V-valued random variables; then A is

(a) certainly bounded if, and only if, it is P-a.s. bounded, *viz.* the random variables of A are P-a.s. uniformly bounded;

(b) perhaps bounded if, and only if, for every $\varepsilon > 0$, there is a ball B_ε in $(V, \|\cdot\|)$ such that all the random variables in A take values in B_ε with probability greater than $1 - \varepsilon$;

(c) perhaps unbounded if, and only if, there exists $]0,1[$ such that, for every t in $]0,+\infty[$, there is a random variable p in A such that

$$P\{\omega \in \Omega : \|p(\omega)\| \geq t\} \geq \beta > 0;$$

in other words, with strictly positive probability, the radius of A is actually infinite;

(d) certainly unbounded if, and only if, for every $\varepsilon > 0$ and for every $t \in]0,+\infty[$, there is a random variable p in A such that

$$P\{\omega \in \Omega : \|p(\omega)\| \geq t\} > 1 - \varepsilon.$$

We shall now examine the behavior of subsets in the space of Example 2.2.1 and in the canonical PN space associated with a normed space $(V, \|\cdot\|)$.

Example 6.2.5. Let A be a bounded subset of $\mathbb{R}$ and set $H := \sup_{p\in A} |p|$. Then, in the PN space $(\mathbb{R}, \nu, \tau_\Pi, \tau_{\Pi^*})$ of Example 2.2.1, one has, for every $t \in]0,+\infty[$,

$$R_A(t) = exp(-H),$$

which belongs to $]0,1[$; thus A is perhaps bounded. The only certainly bounded set in this PN space is the singleton $\{0\}$.

If $A \subset \mathbb{R}$ is unbounded, then for every $K > 0$ there exists $p \in A$ such that $|p| > K$, so that for every $t \in]0,+\infty[$, one has $R_A(t) = 0$, i.e., $R_A = \varepsilon_\infty$; therefore A is certainly unbounded.

Example 6.2.6. Let A be a bounded set in the normed space $(V, \|\cdot\|)$ and set $H := \sup_{p\in A} \|p\| < +\infty$. Then, for $t \in]0,+\infty[$, one has

$$R_A(t) \geq \frac{t}{t+H}.$$

Therefore A is $\mathcal{D}$-bounded; more precisely, it is perhaps bounded if $A \neq 0$, while it is certainly bounded if $A = 0$. An argument similar to that of the previous example shows that, if A is unbounded in $(V, \|\cdot\|)$, then it is certainly unbounded in the canonical PN space associated with it.

If A is a $\mathcal{D}$-bounded set in the PN space (V, ν, τ, τ^*) then the set αA, where α is a real number, need not be $\mathcal{D}$-bounded. The following theorem provides a sufficient condition for this to happen.

Theorem 6.2.3. *Let A be a $\mathcal{D}$-bounded set in the PN space (V, ν, τ, τ^*). If $\mathcal{D}^+$ is stable with respect to τ, i.e., if* (1.6.6) *holds, then the set $\alpha A := \{\alpha p : p \in A\}$ is also $\mathcal{D}$-bounded for every fixed real number α.*

Proof. Because of condition (N2) it suffices to consider the case $\alpha \geq 0$. If either $\alpha = 0$ or $\alpha = 1$ holds, then the assertion is true. Now let α belong to the open interval $]0, 1[$. Then, for every $p \in A$, one has $\nu_{\alpha p} \geq \nu_p$ (see Lemma 3.3.1). Therefore, one has

$$\nu_{\alpha p} \geq \nu_p \geq R_A,$$

which means that αA is $\mathcal{D}$-bounded.

If $\alpha > 1$, let $k = [\alpha] + 1$; again by Lemma 3.3.1, one has $\nu_{\alpha p} \geq \nu_{kp}$. The function G_α defined on $\mathcal{D}^+ \times \cdots \times \mathcal{D}^+$ (k factors) via

$$G_\alpha := \tau^{k-1}(R_A, R_A, \ldots, R_A)$$

belongs to $\mathcal{D}^+$. It follows by induction that

$$\begin{aligned}\nu_{kp} &\geq \tau(\nu_{(k-1)p}, \nu_p) \geq \tau(\tau(\nu_{(k-2)p}, \nu_p), \nu_p) \geq \cdots \geq \\ &\geq \tau^{k-1}(\nu_p, \nu_p, \ldots, \nu_p) \geq \tau^{k-1}(R_A, R_A, \ldots, R_A) = G_\alpha.\end{aligned}$$

As a consequence, one has, for every $p \in A$,

$$\nu_{\alpha p} \geq \nu_{kp} \geq G_\alpha,$$

so that $R_{\alpha A} \geq G_\alpha$, which concludes the proof. □

The reader should bear in mind Theorem 1.6.2 for situations in which the sufficient condition of the previous (and the next) theorem holds.

Theorem 6.2.4. *Let A and B be $\mathcal{D}$-bounded sets in the PN space (V, ν, τ, τ^*); then their union $A \cup B$ is also $\mathcal{D}$-bounded, whenever the set $\mathcal{D}^+$ is stable under τ, i.e. $\tau(\mathcal{D}^+ \times \mathcal{D}^+) \subset \mathcal{D}^+$), namely when condition* (1.6.6) *holds.*

Proof. Equations (6.1.1) and (6.1.2) state respectively

$$R_{A \cup B} \geq \tau(R_A, R_B), \quad \text{and} \quad R_{A+B} \geq \tau(R_A, R_B)$$

and the assertion follows from the fact both R_A and R_B belong to $\mathcal{D}^+$ and that $\mathcal{D}^+$ is stable under τ. □

Notice that the intersection $A \cap B$ of two $\mathcal{D}$-bounded sets A and B is always $\mathcal{D}$-bounded. Indeed, by the definition of the probabilistic radius, one has both $R_{A \cap B} \geq R_A$ and $R_{A \cap B} \geq R_B$. Actually, it suffices that just one of the two sets is $\mathcal{D}$-bounded for the intersection $A \cap B$ to be $\mathcal{D}$-bounded, since

$$R_{A \cap B} \geq \max\{R_A, R_B\}.$$

Given a PN space (V, ν, τ, τ^*), let $\mathcal{P}_{\mathcal{D}^+}(V)$ denote the set of its $\mathcal{D}$-bounded sets. If the stability condition (1.6.6) holds, then $\mathcal{P}_{\mathcal{D}^+}(V)$ is a linear space. This is an immediate consequence of Theorems 6.2.3 and 6.2.4. We state it as a theorem.

Theorem 6.2.5. *If* (V, ν, τ, τ^*) *is a PN space, and if condition* (1.6.3) *holds, then the triple* $(\mathcal{P}_{\mathcal{D}^+}(V), +, \cdot)$ *is a real linear space.*

Definition 6.2.2. We say that probabilistic norm $\nu : \mathbb{R} \to \Delta^+$ has the Lafuerza-Guillén property (LG-property) if, for every $x > 0$, $\lim_{p\to\infty} \nu_p(x) = 0$, or equivalently, $\lim_{p\to\infty} \nu_p = \varepsilon_\infty$.

Example 6.2.7. The quadruple $(\mathbb{R}, \nu, \tau_\Pi, \tau_{\Pi^*})$ where $\nu : \mathbb{R} \to \Delta^+$ is defined by

$$\nu_p(x) := \begin{cases} 0, & x = 0, \\ exp(-|p|^{1/2}), & x \in]0, +\infty[, \\ 1, & x = +\infty \end{cases}$$

where $\nu_0 = \varepsilon_0$ is a PN space but is not a Šerstnev space, and the probabilistic norm has the LG-property.

Example 6.2.8. We consider the PN space $(\mathbb{R}, \nu, \tau, \mathbf{M})$, where τ is a triangle function such that $\tau(\varepsilon_c, \varepsilon_d) \leq \varepsilon_{c+d}$, $\mathbf{M}$ is the maximal triangle function, and the probabilistic norm $\nu : \mathbb{R} \to \Delta^+$ is defined by

$$\nu_p := \varepsilon_{\frac{|p|}{a+|p|}}$$

for every p in $\mathbb{R}$ and for a fixed $a > 0$, with $\nu_p(+\infty) = 1$. With this norm, $\mathbb{R}$ is $\mathcal{D}$-bounded because $\nu_p \geq \varepsilon_1$ for all $p \in \mathbb{R}$. This example does not have the LG-property.

Lemma 6.2.2. *In a PN space* $(\mathbb{R}, \nu, \tau, \tau^*)$ *in which the probabilistic norm has the LG-property, if* $A \subset \mathbb{R}$ *is* $\mathcal{D}$*-bounded then it is classically bounded.*

Proof. If A is $\mathcal{D}$-bounded, there exists a *d.f.* $G \in \mathcal{D}^+$ such that $\nu_p \geq G$ for every $p \in A$, but if A is not classically bounded, then for every $k > 0$ there exists a $p \in A$ such that $|p| > k$. By hypothesis we have $\lim_{p\to\infty} \nu_p(x) = 0$ and therefore for every $x \in]0, +\infty[$ one has $G(x) = 0$, which is a contradiction. □

The converse of the above lemma is, in general, not true. (See Example 6.2.7, in which the only $\mathcal{D}$-bounded set is the singleton $\{0\}$.)

It is now possible to study the $\mathcal{D}$-boundedness of the τ_1-product of two PN spaces under the *same* pair of triangle functions; let these spaces be $(V_1, \nu_1, \tau, \tau^*)$ and $(V_2, \nu_2, \tau, \tau^*)$ (see Section 5.1).

Theorem 6.2.6. *Let $(V_1, \nu_1, \tau, \tau^*)$ and $(V_2, \nu_2, \tau, \tau^*)$ be two PN spaces under the same pair of triangle functions τ and τ^*, let τ_1 be a triangle function such that $\tau^* \gg \tau_1$ and $\tau_1 \gg \tau$, and let $(V_1 \times V_2, \nu_1\tau_1\nu_2, \tau, \tau^*)$ be their τ_1-product. If $\mathcal{D}^+$ is stable under τ_1, $\tau_1(\mathcal{D}^+, \mathcal{D}^+) \subset \mathcal{D}^+$ and if A and B are $\mathcal{D}$-bounded subsets of the PN spaces $(V_1, \nu_1, \tau, \tau^*)$ and $(V_2, \nu_2, \tau, \tau^*)$, respectively, then their Cartesian product $A \times B$ is a $\mathcal{D}$-bounded subset of $(V_1 \times V_2, \nu_1\tau_1\nu_2, \tau, \tau^*)$.*

Proof. It follows from the isotony of τ_1 that

$$\begin{aligned}\inf\{(\nu_1\tau_1\nu_2)(p,q) : p \in A, q \in B\} &= \inf\{\tau_1(\nu_1(p), \nu_2(q)) : p \in A, q \in B\} \\ &= \tau_1(R_A, R_B).\end{aligned}$$

Therefore $R_{A\times B} = \tau_1(R_A, R_B)$, which belongs to $\mathcal{D}^+$. □

We recall that a subset $A \subset V$ of a uniform space $(V, \mathbf{U})$ is said to be uniformly bounded if, and only if, for every circled neighborhood U of the origin θ, there exists a natural number $k \in \mathbb{N}$ such that $A \subset kU$.

The following result shows that, in a Šerstnev PN space, uniform boundedness (named *boundedness* in the present setting) of a subset $A \subset V$ with respect to the strong topology is equivalent to the fact that the probabilistic radius R_A of A is an element of $\mathcal{D}^+$.

Theorem 6.2.7. *For a subset A of a Šerstnev space (V, ν, τ) the following statements are equivalent:*

(a) *A is bounded;*
(b) *R_A belongs to $\mathcal{D}^+$.*

Proof. *(a)* ⇒ *(b)*. Let A be bounded and consider the neighborhood of the origin θ, $\mathrm{N}_\theta(1/n)$. Then, there exists $k \in \mathbb{N}$ such that, for every $p \in A$, one has $p = kq$ for some q in $\mathrm{N}_\theta(1/n)$. Take $t > k/n$, then, because of $(\check{S})$, one has

$$\nu_p(t) = \nu_{kq}(t) = \nu_q\left(\frac{t}{k}\right) \geq \nu_q\left(\frac{1}{n}\right) > 1 - \frac{1}{n},$$

so that

$$R_A(t) \geq 1 - \frac{1}{n},$$

namely $R_A \in \mathcal{D}^+$.

$(b) \Rightarrow (a)$ Let R_A belong $\mathcal{D}^+$. Then for every $n \in \mathbb{N}$, there exists $t_n > 0$ such that $R_A(t_n) > 1 - 1/n$. Therefore, for every $p \in A$, one has

$$\nu_p(t_n) \geq R_A(t_n) > 1 - \frac{1}{n}.$$

Set $k := \min\{j \in \mathbf{N} : j/n \geq t_n\}$. Then

$$\nu_p\left(\frac{k}{n}\right) \geq \nu_p(t_n) > 1 - \frac{1}{n}.$$

Now property $(\check{S})$ yields $\nu_{p/k}(\frac{1}{n}) > 1 - \frac{1}{n}$; in other words, p/k belongs to $\mathbb{N}_\theta(1/n)$, *viz.* there exists $q \in \mathbb{N}_\theta(1/n)$ such that $p = kq$; this means

$$A \subset k\mathbb{N}_\theta(1/n),$$

so that A is bounded. □

As mentioned in Section 3.3, if the triangle function τ^* is Archimedean, i.e., if τ^* admits no idempotents other than ε_0 and ε_∞, then the mapping (6.2.3) is continuous and, as a consequence, the PN space (V, ν, τ, τ^*) is a TV space. If the requirement that τ^* be Archimedean is dropped, then (V, ν, τ, τ^*) need not be a TV space; in this latter case the condition characterizing boundedness takes a more complicated form (see, for instance, p. 130 in James (1990)). But even when τ^* is Archimedean, the present state of our knowledge about PN spaces prevents us from deciding, one way or the other, whether a result similar to Theorem 6.2.7 holds. In this regard some new results are presented in the following.

Furthermore, whenever a metric space has the structure of a TV space, another type of boundedness is relevant. A subset A of a TV space E is *topologically bounded* if for every sequence $(\alpha_n) \subset \mathbb{R}$ with $\lim_{n\to+\infty}(\alpha_n) = 0$ and for every sequence $(p_n) \subset A$, then $\lim_{n\to+\infty}(\alpha_n p_n) = \theta$ in the topology of E. As a consequence, if B is a subset of A, we easily know that B is also topologically bounded.

Taking into account Theorems 3.3.1 and 3.3.3 it is possible to paraphrase them in terms of the following theorem.

Theorem 6.2.8. *Every PN space (V, ν, τ, τ^*), when it is endowed with the strong topology induced by the probabilistic norm ν, is a topological vector space if, and only if, for every $p \in V$ the map from $\mathbb{R}$ into V defined by*

$$\lambda \mapsto \lambda p \tag{6.2.3}$$

is continuous at every λ, i.e. for every $\eta > 0$ (we will suppose, without loss of generality, that $\eta \leq 1$), there exists a number $\delta > 0$ such that

$d_S(\nu_{\alpha' p - \alpha p}, \varepsilon_0) < \eta$ whenever $| \alpha' - \alpha |< \delta$; or, equivalently, such that $d_S(\nu_{\beta p}, \varepsilon_0) < \eta$ whenever $| \beta |< \delta$.

The following theorem studies whether certain classes of spaces are topological vector spaces.

Theorem 6.2.9.

(a) *No equilateral space (V, F, Π_M) is a TV space;*
(b) *A Šerstnev PN space is a TV space if, and only if, the probabilistic norm ν maps V into $\mathcal{D}^+$ rather than into Δ^+, viz. $\nu(V) \subseteq \mathcal{D}^+$;*
(c) *A simple space $(V, \|\cdot\|, G, M)$ is a TV space if, and only if, G belongs to $\mathcal{D}^+$;*
(d) *If G is a d.f. different from ε_0 and $\varepsilon_{+\infty}$, then the α-simple space $(V, \|\cdot\|, G; \alpha)$ is a TV space if, and only if, G belongs to $\mathcal{D}^+$;*
(e) *An EN space (S, ν) is a TV space if, and only if, ν_p belongs to $\mathcal{D}^+$ for every $p \in S$.*

Proof. Let θ denote the null vector of the vector space V. Since any PM space and, hence, any PN space, can be metrized, one can limit oneself to investigating the behavior of sequences. Moreover, because of the linear structure of V, one can take $p \neq \theta$ and an arbitrary sequence (λ_n) with $\lambda_n \neq 0 (n \in \mathbb{N})$ such that $\lambda_n \to 0$ as n tends to $+\infty$.

(a) For every $n \in \mathbb{N}$, one has $\nu_{\lambda_n p} = F$ while $\nu_0 = \varepsilon_0$. Therefore the map (6.2.3) is not continuous.
(b) If ν maps V into $\mathcal{D}^+$, then for every $t > 0$, one has

$$\nu_{\lambda_n p}(t) = \nu_p\left(\frac{t}{|\lambda_n|}\right) \xrightarrow[n \to +\infty]{} 1,$$

whence the assertion. Conversely, if there exists at least one $p \in V$ such that $\nu_p \in \Delta^+ \setminus \mathcal{D}^+$, namely such that $\nu_p(x) \xrightarrow[x \to +\infty]{} \gamma < 1$, then, for $x > 0$,

$$\nu_{\lambda_n p}(x) = \nu_p\left(\frac{x}{|\lambda_n|}\right) \xrightarrow[n \to +\infty]{} \gamma < 1,$$

so that the mapping $\lambda \mapsto \lambda p$ is not continuous.
(c) It is a trivial consequence of part (b), since every simple space is a Šerstnev space.

(d) Let (λ_n) be a sequence of real numbers that tends to 0, when n goes to $+\infty$. Then, for all $p \in V$ and $x > 0$, one has, for every $n \in \mathbf{N}$,

$$\nu_{\lambda_n p}(x) = G\left(\frac{x}{\|\lambda_n p\|^\alpha}\right) = G\left(\frac{x}{|\lambda_n|^\alpha \|p\|^\alpha}\right).$$

Therefore $\lim_{n\to+\infty} \nu_{\lambda_n p}(x) = 1$ if, and only if, G belongs to $\mathcal{D}^+$.

(e) The proof is analogous to that of part (b). □

We shall call *strict*[1] any PN space (V, ν, τ, τ^*) such that $\nu(V) \subseteq \mathcal{D}^+$, i.e., such that ν_p belongs to $\mathcal{D}^+$ for every $p \in V$. This definition can be extended to PPN and PSN spaces. Thus, Theorem 6.2.9 (b), (c), (d), and (e) can be rephrased as follows.

Theorem 6.2.10. *Šerstnev spaces, simple spaces, α-simple spaces and EN spaces are TV spaces if, and only if, they are strict.*

Theorem 6.2.11. *Let (V, ν, τ, τ^*) be a PN space in which $\nu(V) \subseteq \mathcal{D}^+$ and $\mathcal{D}^+$ is stable under τ, i.e., $\tau(\mathcal{D}^+ \times \mathcal{D}^+) \subseteq \mathcal{D}^+$. If the sequence $(p_m)_{m\in\mathbf{N}}$ in V converges strongly to $p \in V$ and $A = \{p_m : m \in \mathbf{N}\}$, then A is a $\mathcal{D}$-bounded subset of V.*

Proof. Let $p_m \to p$ in the strong topology of V. Then there exists a positive integer k such that for every $m \geq k$ and for every $G \in \mathcal{D}^+$ one has $\nu_{p_m - p} \geq G$. Therefore

$$\nu_{p_m} \geq \tau(\nu_{p_m - p}, \nu_p) \geq \tau(G, \nu_p).$$

If we define $H := \min\{\nu_{p_1}, \ldots, \nu_{p_{k-1}}, \tau(G, \nu_p)\}$, then $H \in \mathcal{D}^+$ and $\nu_{p_m} \geq H$, for every $m \in \mathbb{N}$. Hence A is a $\mathcal{D}$-bounded set. □

Note that, in Example 6.2.7 in which $\nu(V) \subseteq \Delta^+ \setminus \mathcal{D}^+$, the sequence $(\frac{1}{m})_{m\in\mathbb{N}}$ is convergent but the set $A = \{\frac{1}{m} : m \in \mathbb{N}\}$ is not a $\mathcal{D}$-bounded set.

Theorem 6.2.12. *If the PN space $(\mathbb{R}, \nu, \tau, \tau^*)$ is a TV space then it is complete.*

---

[1]For every PN space (V, ν, τ, τ^*), if $p \in V$ and $x \geq 0$, then $\nu_p(x)$ may be thought of as the probability $P(\|p\| < x)$, where $\|\cdot\|$ is a norm for V. So the fact that ν_p does not belong to $\mathcal{D}^+$ means that $P(\|p\| < +\infty) < 1$; this is to be regarded as being "odd".

Proof. Let (p_m) be a strong Cauchy sequence, for every $m, n \in \mathbb{N}$, and $m > n$, one has

$$\lim_{m,n\to\infty} \nu_{p_m - p_n} = \varepsilon_0.$$

Because of the PN space $(\mathbb{R}, \nu, \tau, \tau^*)$ is a TV space by hypothesis, one has

$$\nu_{\lim_{m,n\to\infty}(p_m - p_n)} = \varepsilon_0 = \nu_0.$$

Hence (p_m) is a classical Cauchy sequence in $\mathbb{R}$; therefore, it is convergent to $p \in \mathbb{R}$, i.e., $\lim_{m\to\infty} p_m = p$, and since the PN space given is TV space one has $\lim_{m\to\infty} \nu_{p_m - p} = \nu_0 = \varepsilon_0$, and the proof is complete. □

However, regarding general PN spaces, the condition $\nu(V) \subseteq \mathcal{D}^+$ is not necessary to obtain a TV space: see Theorem 7.2.2.

Definition 6.2.3. The PN space (V, ν, τ, τ^*) is said to satisfy the *Double infinity condition* (*DI-condition*) if the probabilistic norm ν is such that, for all $\alpha \in \mathbb{R} \setminus \{0\}$, $x \in \mathbb{R}$ and $p \in V$,

$$\nu_{\alpha p}(x) = \nu_p(\varphi(\alpha, x)),$$

where $\varphi : \mathbb{R} \times [0, +\infty[\to [0, +\infty[$ satisfies

$$\lim_{x\to+\infty} \varphi(\alpha, x) = +\infty \quad \text{and} \quad \lim_{\alpha\to 0} \varphi(\alpha, x) = +\infty.$$

Example 6.2.9. If (V, ν, τ, τ^*) is a Šerstnev space then it satisfies the DI-condition. It is sufficient to check that in Šerstnev spaces $\varphi(\alpha, x) = \frac{x}{|\alpha|}$.

Example 6.2.10. Let $(V, \|\cdot\|)$ be a normed space. For $\beta \in]0, 1[$, define $\nu : V \to \Delta^+$ by

$$\nu_p(x) := \begin{cases} 0, & x \le 0, \\ (1 - \beta\,\varepsilon_0(\|p\|))\,\dfrac{\ln(1+x)}{\ln(1+x) + \|p\|}, & x \in]0, +\infty[, \\ 1, & x = +\infty. \end{cases}$$

Below, we shall prove that $(V, \nu, \tau_\Pi, \tau_M)$

(1) is a PN space;
(2) is neither a Šerstnev space nor a TV space nor a strict PN space;
(3) satisfies the DI-condition, with

$$\varphi(\alpha, x) = (x+1)^{1/|\alpha|} - 1.$$

(1) $(V, \nu, \tau_\Pi, \tau_M)$ is a PN space. It is easily ascertained that (N1) and (N2) hold.

For all p and q in V, for all s and t in R^+, one has

$$\frac{\|p+q\|}{\ln(1+s+t)} \le \frac{\|p\|+\|q\|}{\ln(1+s+t)} \le \frac{\|p\|}{\ln(1+s)} + \frac{\|q\|}{\ln(1+t)}.$$

Notice that the factor $(1-\beta\,\varepsilon_0(\|p\|))$ equals 1 if, and only if, $p=\theta$; otherwise, it equals $1-\beta$. Then, if both p and q differ from θ,

$$\begin{aligned}
\nu_{p+q}(s+t) &= (1-\beta)\,\frac{\ln(1+s+t)}{\ln(1+s+t)+\|p+q\|} = (1-\beta)\,\frac{1}{1+\frac{\|p+q\|}{\ln(1+s+t)}} \\
&\ge (1-\beta)\,\frac{1}{1+\frac{\|p\|}{\ln(1+s)}+\frac{\|q\|}{\ln(1+t)}} \\
&\ge (1-\beta)^2\,\frac{1}{1+\frac{\|p\|}{\ln(1+s)}+\frac{\|q\|}{\ln(1+t)}+\frac{\|p\|\,\|q\|}{\ln(1+s)\,\ln(1+t)}} \\
&= \frac{(1-\beta)}{1+\frac{\|p\|}{\ln(1+s)}}\cdot\frac{(1-\beta)}{1+\frac{\|q\|}{\ln(1+t)}} \\
&= (1-\beta)\,\frac{\ln(1+s)}{\ln(1+s)+\|p\|}\cdot(1-\beta)\,\frac{\ln(1+t)}{\ln(1+t)+\|q\|} \\
&= \nu_p(s)\cdot\nu_q(t).
\end{aligned}$$

As a consequence (N3) holds, i.e., for all p and q in V, and for every $x \in \mathbb{R}^+$,

$$\nu_{p+q}(x) \ge \tau_\pi(\nu_p,\nu_q)(x).$$

This latter inequality is trivially true if at least one of p and q equals θ.

For all s and t in $\mathbb{R}^+$, one has

$$\begin{aligned}
1-\frac{\ln(1+s)}{\ln(1+s+t)} &= \frac{\ln(1+s+t)-\ln(1+s)}{\ln(1+s+t)} = \frac{\ln\frac{s+t+1}{s+1}}{\ln(1+s+t)} \\
&= \frac{\ln(1+\frac{t}{s+1})}{\ln(1+s+t)} \le \frac{\ln(1+t)}{\ln(1+s+t)}.
\end{aligned}$$

Therefore, because of Lemma 3.3.1,

$$\nu_{\left(1-\frac{\ln(1+s)}{\ln(1+s+t)}\right)p} \ge \nu_{\frac{\ln(1+t)}{\ln(1+s+t)}\,p}.$$

For all $\alpha \in\,]0,1[$ and $x \in \mathbb{R}^+$, there exist s_0, t_0 in R^+ such that $s_0 + t_0 = x$ and

$$\alpha = \frac{\ln(1+s_0)}{\ln(1+s_0+t_0)};$$

in fact $s_0 = (1+x)^\alpha - 1$. Whence, if $p \neq \theta$,

$$\begin{aligned}
\tau_M(\nu_{\alpha p}, \nu_{(1-\alpha)p})(x) &= \sup_{s+t=x} \min(\nu_{\alpha p}(s), \nu_{(1-\alpha)p}(t)) \\
&= \sup_{s+t=x} \min\left(\nu_{\left(\frac{\ln(1+s_0)}{\ln(1+s_0+t_0)}\right)p}(s), \nu_{\left(1-\frac{\ln(1+s_0)}{\ln(1+s_0+t_0)}\right)p}(t)\right) \\
&\geq \min\left(\nu_{\left(\frac{\ln(1+s_0)}{\ln(1+s_0+t_0)}\right)p}(s_0), \nu_{\left(1-\frac{\ln(1+s_0)}{\ln(1+s_0+t_0)}\right)p}(t_0)\right) \\
&\geq \min\left(\nu_{\left(\frac{\ln(1+s_0)}{\ln(1+s_0+t_0)}\right)p}(s_0), \nu_{\left(\frac{\ln(t_0+1)}{\ln(1+s_0+t_0)}\right)p}(t_0)\right) \\
&= \min\left(\frac{(1-\beta)\,\ln(1+s_0+t_0)}{\ln(1+s_0+t_0)+\|p\|},\right. \\
&\qquad \left.\frac{(1-\beta)\,\ln(1+s_0+t_0)}{\ln(1+s_0+t_0)+\|p\|}\right) \\
&= \min(\nu_p(x), \nu_p(x)) = \nu_p(x).
\end{aligned}$$

Therefore (N4) holds for $p \neq \theta$; since (N4) is obvious for $p = \theta$, it holds for every $p \in V$ and for every $\alpha \in\,]0,1[$.

For $\alpha = 0$ or $\alpha = 1$, it is obvious that (N4) holds, i.e.,

$$\nu_p = \tau_M(\nu_{\alpha p}, \nu_{(1-\alpha)p}).$$

(2) $(V, \nu, \tau_\Pi, \tau_M)$ is neither a Šerstnev space nor a TV space nor a strict PN space. A straightforward calculation shows that $\nu_{\alpha p}(x) \neq \nu_p(x/|\alpha|)$ so that $(V, \nu, \tau_\pi, \tau_M)$ is not a Šerstnev space.

We recall that a sufficient condition for a PN space (V, ν, τ, τ^*) to be a TV space is that τ^* be an Archimedean triangle function. Now, τ_M is not Archimedean on all of Δ^+ so that all one needs to prove is that the scalar multiplication $\lambda \mapsto \lambda p$ is not continuous. For every sequence (λ_n) of real numbers that converges to 0 as n tends to $+\infty$, and for every p in V one has $\lambda_n\, p \nrightarrow \theta$ in the strong topology of V. Now

$$\lim_{n\to\infty} \nu_{\lambda_n p}(x) = 1 - \beta < 1,$$

or equivalently,

$$\lim_{n\to\infty} \nu_{\lambda_n\, p} \neq \varepsilon_0.$$

Also since $\lim_{x\to\infty} \nu_p(x) = 1 - \beta < 1$, the distance d.f. ν_p is in $\Delta^+ \setminus \mathcal{D}^+$, so that $(V, \nu, \tau_\pi, \tau_M)$ is not a strict PN space.

(3) $(V, \nu, \tau_\Pi, \tau_M)$ satisfies the DI-condition. For all $x \in \mathbb{R}^+$, $\lambda \in \mathbb{R} \setminus \{0\}$ and $p \in V$, one has, for $p \neq \theta$,

$$\nu_{\lambda\, p}(x) = \frac{(1-\beta)\,\ln(1+x)}{\ln(1+x) + |\lambda|\,\|p\,\|} = \frac{(1-\beta)\,\frac{1}{|\lambda|}\ln(1+x)}{\frac{1}{|\lambda|}\ln(1+x) + \|p\,\|}$$

$$= \frac{(1-\beta)\,\ln(1+x)^{\frac{1}{|\lambda|}}}{\ln(1+x)^{\frac{1}{|\lambda|}} + \|p\,\|} = \nu_p((1+x)^{1/|\lambda|} - 1).$$

Let

$$\varphi(\lambda, x) = (1+x)^{1/|\lambda|} - 1.$$

It is easily checked that $(\lambda, x) \mapsto \varphi(\lambda, x)$ satisfies

$$\lim_{x\to+\infty} \varphi(\lambda, x) = +\infty \quad \text{and} \quad \lim_{\lambda\to 0} \varphi(\lambda, x) = +\infty.$$

Theorem 6.2.13. *Let (V, ν, τ, τ^*) be a PN space which satisfies the DI-condition. Then for a subset $A \subset V$ the following statements are equivalent:*

(a) *A is $\mathcal{D}$-bounded.*
(b) *A is bounded, namely, for every $n \in \mathbb{N}$ and for every $p \in A$, there is $k \in \mathbb{N}$ such that $\nu_{p/k}(1/n) > 1 - 1/n$.*
(c) *A is topologically bounded.*

Proof. (a) $\Longrightarrow$ (b). Let A be any $\mathcal{D}$-bounded subset of V. By definition of $\mathcal{D}$-boundedness, the probabilistic radius R_A of A is a distance d.f. such that $\lim_{x\to+\infty} R_A(x) = 1$, Therefore, for every $n \in \mathbb{N}$, there exists $x_n > 0$ such that $R_A(x_n) > 1 - 1/n$. Thus, for every $p \in A$,

$$\nu_p(x_n) \geq R_A(x_n) > 1 - 1/n.$$

Since, in view of the φ-condition,

$$\lim_{\lambda\to 0} \varphi(\lambda, 1/n) = +\infty,$$

for every $n \in \mathbb{N}$, there exists $\lambda' \in \mathbb{R}$ such that $\varphi(\lambda', 1/n) \geq x_n$. Therefore,

$$\nu_{\lambda' p}(1/n) = \nu_p(\varphi(\lambda', 1/n)) \geq \nu_p(x_n) \geq R_A(x_n) > 1 - 1/n.$$

As a consequence of letting $k = [1/\lambda']$, where $[t]$ denotes the integral part of t, one has

$$\nu_{p/k}(1/n) > 1 - 1/n,$$

namely, A is bounded.
(b) $\Longrightarrow$ (a) Let A be a bounded subset of V and consider the neighbourhood of θ, $N_\theta(1/n)$. Then there exists $\lambda_0 \in \mathbb{R}$ such that for every $p \in V$, $p = \lambda_0\, q$ for some $q \in N_\theta(1/n)$.

Because of the DI-condition, $\lim_{x\to+\infty} \varphi(\lambda_0, x) = +\infty$, for every $n \in \mathbb{N}$; then, there exists $x_0 > 0$ such that $\varphi(\lambda_0, x_0) \geq 1$. Then, for $x \geq x_0$,

$$\nu_p(x) = \nu_{\lambda_0\, q}(x) \geq \nu_{\lambda_0\, q}(x_0) = \nu_q(\varphi(\lambda_0, x_0)) \geq \nu_q(1) \geq \nu_q(1/n) > 1 - 1/n,$$

so that

$$R_A(x) \geq 1 - \frac{1}{n},$$

i.e., R_A is in $\mathcal{D}^+$. (a)$\Longrightarrow$ (c). Let A be any $\mathcal{D}$-bounded subset of V. One has

$$\nu_{\alpha_n p_n}(x) = \nu_{p_n}(\varphi(\alpha_n, x)) \geq R_A \to 1,$$

which means that $\alpha_n p_n \to \theta$ whenever $n \to +\infty$.
(c)$\Longrightarrow$ (a). Let A be a subset of V which is not $\mathcal{D}$-bounded. Then

$$\lim_{x\to+\infty} R_A(x) = \gamma < 1.$$

By definition of R_A, for every $n \in \mathbb{N}$ there is $p_n \in A$ such that

$$\nu_{p_n}(\varphi(\alpha_n, x) < \frac{1+\lambda}{2} < 1.$$

Then for every $x \in [0, n]$,

$$\nu_{\alpha_n p_n}(x) = \nu_{p_n}(\varphi(\alpha_n, x)) < \frac{1+\lambda}{2} < 1,$$

which shows that $\nu_{\alpha_n p_n}$ does not tend to ε_0, even if it has a weak limit, *viz.* $\{\alpha_n p_n\}$ does not tend to θ in the strong topology; in other words, A is not topologically bounded. □

The next example exhibits a PN space (V, ν, τ, τ^*) that is a TV space but in which the triangle function τ^* is not Archimedean.

Example 6.2.11. Let (V, ν, τ, τ^*) be a PN space with τ^* a non Archimedean triangle function. Let f, g be functions satisfying the following conditions:

(i) f is continuous non-increasing from $[0, +\infty]$ into [0,1] and $f(0) = 1$;
(ii) g is a map from $\mathbb{R}^+ \times \mathbb{R}^+$ into [0,1], continuous in the first and in the second places, non-decreasing in the first place and non-increasing in the second place with $g(x, 0) = 1$. Then the quadruple (V, ν, τ, τ^*) where the probabilistic norm ν is defined via

$$\nu_p(x) := f(\|p\|) \cdot g(x, \|p\|),$$

is a TV space.

Notice that every PN space (V, ν, τ, τ^*) in which $\tau^* = \tau_M$ such that $\nu(V) \subseteq \mathcal{D}^+$ is a TV space, since τ_M is Archimedean on the subset $\mathcal{D}^+$ of Δ^+.

The condition $\nu(V) \subseteq \mathcal{D}^+$ is not necessary to obtain a TV space, as shown in the following example.

Example 6.2.12. With the same assumptions of Example 6.2.8 one has the quadruple $(V, \nu, \prod_{\Pi}, \prod_M)$ where

$$\nu_p(x) = \frac{1}{1 + \|p\|} \cdot \frac{x}{x + \|p\|}$$

is a PN space that is a TV space, and is not a Šerstnev PN space. The PN space $(V, \nu, \prod_{\Pi}, \prod_M)$ is not strict.

Example 6.2.13. The quadruple $(V, \nu, \prod_{\Pi}, \prod_M)$ where ν, the probabilistic norm, is a map $\nu : V \to \Delta^+$ defined via $\nu_p(x) := e^{-\frac{\|p\| x}{\|p\| x + 1}}$, is a PN space that is neither a Šerstnev PN space nor strict, and is a TV space.

Example 6.2.14. For $p \in \mathbb{R}$ define ν by setting $\nu_0 = \varepsilon_0$, and

$$\nu_p := \frac{1}{\mid p \mid +2} \varepsilon_0 + \frac{\mid p \mid +1}{\mid p \mid +2} \varepsilon_\infty.$$

For $p \neq 0$, it is straightforward to show that $(\mathbb{R}, \nu, \tau_W, \tau_M)$ is a PN space, and that, for any $p \neq 0$ and any sequence (α_n) converging to 0, the sequence $(\nu_{\alpha_n p})$ converges weakly, not to ε_0, but to $\frac{1}{2}(\varepsilon_0 + \varepsilon_\infty)$. Thus $\nu_{\alpha p}$ is not continuous in its first place, i.e., for a fixed $p \in V$, scalar multiplication is not a continuous mapping from $\mathbb{R}$ into V. Thus $(\mathbb{R}, \nu, \tau_W, \tau_M)$ is not a TV space.

Example 6.2.15. Let $(V, \|\cdot\|)$ be a normed space and, for $\alpha \in \,]0,1[$, let $\nu : V \to \Delta^+$ be given by

$$\nu_p(x) = \begin{cases} 0, & x \leq 0, \\ \dfrac{\ln(1+x)}{\ln(1+x) + \|p\|}, & x \in \,]0, +\infty[, \ \ \|p\| < 1 \\ \dfrac{a\ln(1+x)}{\ln(1+x) + \|p\|}, & x \in \,]0, +\infty[, \ \ \|p\| \geq 1, \\ 1, & x = +\infty. \end{cases}$$

Then

(1) $(V, \nu, \tau_\pi, \tau_M)$ is a PN space satisfying the DI-condition with

$$\varphi(\lambda, x) = (1+x)^{\frac{1}{\|\lambda\|}} - 1;$$

(2) $(V, \nu, \tau_\pi, \tau_M)$ is a TV PN space;
(3) the subset $A = \{p : \|p\| < 1\}$ is both $\mathcal{D}$-bounded and bounded.

Only property (2) needs to be checked. For every sequence (λ_n) of real numbers that converges to 0 as n tends to $+\infty$, and for every $p \in V$, one has $\lambda_n p \to \theta$ in the strong topology of V; in fact, for every $x \in \mathbb{R}^+$ $\lim_{n\to\infty} \nu_{\lambda_n p}(x) = 1$, namely, $\lim_{n\to\infty} \nu_{\lambda_n p} = \varepsilon_0$.

The following theorem studies whether certain classes of PN spaces are TV spaces.

We shall now pass on to address the problem of Archimedeanity of triangle functions τ^* of the type $\tau_{T,L}$. As a consequence of this a class of PN spaces that are TV spaces is given.

Definition 6.2.4. Let $\mathcal{L}$ be the set of all binary operations L on $\mathbb{R}^+$ satisfying the following conditions:

(i) $RanL = \mathbb{R}^+$;
(ii) L is non-decreasing in each place;
(iii) L is continuous on $\mathbb{R}^+ \times \mathbb{R}^+$ except, at most, in $(0, \infty)$ and $(\infty, 0)$.

Theorem 6.2.14. *Let T be a left-continuous t-norm, and let L belong to $\mathcal{L}$ and satisfy the following conditions:*

(a) *L is commutative;*
(b) *L is associative;*

(c) $u_1 < u_2$ *and* $v_1 < v_2$ *imply* $L(u_1, v_1) < L(u_2, v_2)$;
(d) $L(x, 0) = x$;

then the function $\tau_{T,L}$ *is a triangle function.*

Lemma 6.2.3. *Let* $L \in \mathcal{L}$ *satisfy the conditions of Theorem* 6.2.10. *Then the following statements hold:*

(i) $L \geq \max$,
(ii) *For every* $a \in]0, +\infty[$ *one has* $L = \max \Leftrightarrow L(a, a) = a$.

Lemma 6.2.4. *Let* T *be a continuous t-norm and let* $L \in \mathcal{L}$ *satisfy the condition of Theorem* 6.2.10 (a); *then* $\tau_{T,L}(\varepsilon_a, \varepsilon_b) = \varepsilon_{L(a,b)}$ *for all* a *and* b *in* $[0, +\infty[$.

Theorem 6.2.15. *Let* T *be an Archimedean t-norm and assume that* $L \in \mathcal{L}$ *satisfies the assumptions of Theorem* 6.2.10, *then the following statements are equivalent:*

(a) $\tau_{T,L}$ *is a triangle function having no non-trivial idempotents in* Δ^+;
(b) L *satisfies the condition* $L(x, x) > x$ *for every* $x \in]0, +\infty[$.

Proof. According to Theorems 7.2.4 and 7.2.8 in [15] the function $\tau_{T,L}$ is a continuous triangle function.

(a) $\Rightarrow$ *(b)* A continuous triangle function with no non-trivial idempotents in Δ^+ is Archimedean. Then $\tau_{T,L}$ is an Archimedean triangle function, and, for every function $F \notin \{\varepsilon_0, \varepsilon_\infty\}$ one has $\tau_{T,L}(F, F) \neq F$. By Lemma 1.1 $\varepsilon_a \neq \tau_{T,L}(\varepsilon_a, \varepsilon_a) = \varepsilon_{L(a,a)}$ for every $a \in]0, +\infty[$. Therefore $L(a, a) \neq a$ and from Lemma 6.2.2 one has $L(a, a) > a$, which is the assertion.

(b) $\Rightarrow$ *(a)* Let $F \in \Delta^+$ be different from ε_a for every $a \in]0, +\infty[$. Then there exists $x_0 \in]0, +\infty[$ such that $F(x_0)$ is in $]0, 1[$. Since T is Archimedean, the following holds

$$\begin{aligned}\tau_{T,L}(F, F)(x_0) &= \sup\{T(F(u), F(v)) : L(u, v) = x_0\} \\ &\leq T(F(x_0), F(x_0)) < F(x_0).\end{aligned}$$

Therefore F is not an idempotent element of $\tau_{T,L}$. Now let a be in $]0, +\infty[$. By Lemma 6.2.3 $\tau_{T,L}(\varepsilon_a, \varepsilon_a) = \varepsilon_{L(a,a)}$ and by hypothesis (b) one has $\varepsilon_{L(a,a)} \neq \varepsilon_a$, whence $\tau_{T,L}(\varepsilon_a, \varepsilon_a) \neq \varepsilon_a$ and $\tau_{T,L}$ has no non-trivial idempotents elements. This concludes the proof. $\square$

Example 6.2.16. Copulas of the form (1.4.2) are called *Archimedean*. An Archimedean copula is associative and a t-norm. As a consequence, the triangle functions $\tau_{C,L}$ where C is an Archimedean copula and $L \in \mathcal{L}$ satisfies the assumptions of Theorem 6.2.10, are Archimedean.

Theorem 6.2.16. *Let T and C be a continuous t-norm and an Archimedean copula, respectively; if, for every $(u,v) \in [0,1]^2$, $T(u,v) \leq C(u,v)$, then every PN space of the form $(V, \nu, \tau_{T,L}, \tau_{C,L})$ is a TV space.*

The above result provides many examples of Archimedean triangle functions. It suffices to take

$$L = K_\alpha(x,y) = (x^\alpha + y^\alpha)^{1/\alpha} \quad (x, y \in \mathbb{R}^+; \alpha \geq 1).$$

Theorem 6.2.17. *Let T_1, T_2 be t-norms such that $T_1 \leq T_2$; and let L_1, L_2 in $\mathcal{L}$ satisfy the assumptions of* Theorem 6.2.10; *and suppose that $L_1 \leq L_2$, then τ_{T_1,L_2} and τ_{T_2,L_1} have the same idempotent elements.*

Proof. If F is in Δ^+ and τ_{T_1,L_2} admits F as idempotent, then

$$F = \tau_{T_1,L_2}(F,F) \leq \tau_{T_2,L_1}(F,F) \leq F,$$

holds and one immediately has $\tau_{T_2,L_1}(F,F) = F$. □

Corollary 6.2.1. *Let T be a t-norm such that $\Pi \leq T$, then the family of all PN spaces of the type $(V, \nu, \tau_{\Pi,L}, \tau_{T,L})$ are TV spaces.*

Corollary 6.2.2. *Let T be a t-norm such that $W \leq T$, then the family of all PN spaces of the type $(V, \nu, \tau_{W,L}, \tau_{T,L})$ are TV spaces.*

6.3 Total Boundedness

We have seen in the previous section that in a particular class of PN spaces the concepts of $\mathcal{D}$-boundedness, boundedness, and topological boundedness are equivalent. Here we prove that other type of boundedness can be considered which are not equivalent to $\mathcal{D}$-boundedness.

The results that follow have been established in Lafuerza-Guillén, Saadati, and Zhang (2012).

Lemma 6.3.1. *Let τ be a continuous triangle function. Then for every $F \in \mathcal{D}^+$ and $F < \varepsilon_0$ there exists $G \geq F$ such that $\tau(G,G) > F$.*

Proof. Let $F \in \mathcal{D}^+$ and $F < \varepsilon_0$ exist such that for every $G \geq F$ one has $\tau(G,G) \leq F$. Consider the sequence of d.f.s defined by

$$G_n = \max\{\varepsilon_{\frac{1}{n}}, F\},$$

then $G_n \geq F$ for every $n \in \mathbb{N}$, therefore $\tau(G_n, G_n) \leq F$. Taking $n \to +\infty$ in the above inequality then one has $\varepsilon_0 \leq F$, which is a contradiction to the hypothesis. □

Definition 6.3.1. Let (V, ν, τ, τ^*) be a PN space and $A \subset V$. A set A is said to be *probabilistic strongly totally bounded* if, and only if, for every $F \in \mathcal{D}^+$ and $F < \varepsilon_0$, there exists a finite subset S_F of A such that

$$A \subseteq \bigcup_{p \in S_F} D_p(F) \tag{6.3.1}$$

where $D_p(F) = \{q \in V : \nu_{p-q} > F\}$.

Lemma 6.3.2. *Let (V, ν, τ, τ^*) be a PN space and $A \subset V$. A is a probabilistic strongly totally bounded set if, and only if, for every $F \in \mathcal{D}^+$ with $F < \varepsilon_0$, there exists a finite subset S_F of V such that*

$$A \subseteq \bigcup_{p \in S_F} D_p(F). \tag{6.3.2}$$

Proof. Let $F \in \mathcal{D}^+$, $F < \varepsilon_0$, and assume that condition (2.2) holds. By continuity of τ, there exists $G \geq F$ such that $\tau(G, G) > F$. Now, applying condition (2.2) for G, there exists a subset $S_G = p_1, \ldots, p_n$ of V such that $A \subseteq \bigcup_{p \in S_G} D_p(G)$. We assume that $D_{p_j}(G) \cap A \neq \emptyset$, otherwise one omits p_j from S_G and so we have $A \subseteq \bigcup_{p_i \in S_G \setminus p_j} D_{p_i}(G)$. For every $i = 1, \ldots, n$ we select q_i in $D_{p_i}(G) \cap A$, and we put $S_F = q_1, \ldots, q_n$. Now for every q in A, there exists $i = 1, \ldots, n$ such that $\nu_{q-p_i} > G$. Therefore one has, by the axiom (N3) in Definition 2.2.1,

$$\nu_{q-q_i} \geq \tau(\nu_{q-p_i}, \nu_{p_i-q_i}) \geq \tau(G, G) > F,$$

which implies that $A \subseteq \bigcup_{q_i \in S_F} D_{q_i}(F)$. The converse is trivial. □

Lemma 6.3.3. *Let (V, ν, τ, τ^*) be a PN space and $A \subset V$. If A is a probabilistic strongly totally bounded set then so is its closure $\overline{A}$.*

Proof. Let $F \in \mathcal{D}^+$, $F < \varepsilon_0$, then there exists a finite subset $S_G = q_1, \ldots, q_n$ of V with $G \geq F$ such that $\tau(G, G) > F$, such that $A \subseteq \bigcup_{q_i \in S_G} D_{q_i}(G)$. Since for every r in $\overline{A}$, $N_r(\frac{1}{n}) \cap A$ is non-empty for every $n \in \mathbb{N}$ it follows that we can find $p \in A$ such that $\nu_{p-r} \geq G$, and there exists $1 \leq i \leq n$ such that $\nu_{p-q_i} \geq G$, therefore $\nu_{r-q_i} \geq \tau(\nu_{r-p}, \nu_{p-q_i}) \geq \tau(G,G) > F$. Hence

$$\overline{A} \subset \bigcup_{q_i \in S_F} D_{q_i}(F),$$

i.e. $\overline{A}$ is a probabilistic strongly totally bounded set. □

Theorem 6.3.1. *Let (V, ν, τ, τ^*) be a PN space and $A \subset V$. A is a probabilistic strongly totally bounded set if, and only if, every sequence in A has a strong Cauchy subsequence.*

Proof. Let A be a probabilistic strongly totally bounded set. Let $(p_n)_{n \in \mathbb{N}}$ be a sequence in A. For every $k \in \mathbb{N}$, there exists a finite subset S_{F_k} of V such that $A \subseteq \bigcup_{q \in S_{F_k}} D_q(F_k)$, with $F_k = \varepsilon_{\frac{1}{k}}$. Hence, for $k = 1$, there exists $q_1 \in S_{F_1}$ and a subsequence $(p_{1,n})$ of (p_n) such that $p_{1,n} \in D_{q_1}(F_1)$, for every $n \in \mathbb{N}$. Similarly, there exists $q_2 \in S_{F_2}$ and a subsequence $(p_{2,n})$ of $(p_{1,n})$ such that $p_{2,n} \in D_{q_2}(F_2)$, for every $n \in \mathbb{N}$. Continuing this process, we get $q_k \in S_{F_k}$ and subsequence $(p_{k,n})$ of $(p_{k-1,n})$ such that $p_{k,n} \in D_{q_k}(F_k)$, for every $n \in \mathbb{N}$. Now we consider the subsequence $(p_{n,n})$ of (p_n). For every $F \in \mathcal{D}^+$ and $F < \varepsilon_0$, by continuity of τ, there exists an $n_0 \in \mathbb{N}$ such that $\tau(F_{n_0}, F_{n_0}) > F$. Therefore for every $k, m \geq n_0$, one has

$$\nu_{p_{k,k}-p_{m,m}} \geq \tau(\nu_{p_{k,k}-q_{n_0}}, \nu_{q_{n_0}-p_{m,m}}) \geq \tau(F_{n_0}, F_{n_0}) > F.$$

Hence $(p_{n,n})$ is a strong Cauchy sequence. Conversely, suppose that A is not a probabilistic strongly totally bounded set. Then there exists $F \in \mathcal{D}^+$ such that for every finite subset S_F of V, A is not a subset of $\bigcup_{q \in S_F} D_q(F)$. Fix $p_1 \in A$. Since A is not a subset of $\bigcup_{q \in p_1} D_q(F)$, there exists $p_2 \in A$ such that $\nu_{p_1-p_2} \leq F$. Since A is not a subset of $\bigcup_{q \in p_1, p_2} D_q(F)$, there exists $p_3 \in A$ such that $\nu_{p_1-p_3} \leq F$ and $\nu_{p_2-p_3} \leq F$. Continuing this process, we construct a sequence (p_n) of distinct points in A such that $\nu_{p_i-p_j} \leq F$, for every $i \neq j$. Therefore (p_n) admits no strong Cauchy subsequences.

Every probabilistic strongly totally bounded set is not a $\mathcal{D}$-bounded set, in general, as you can see from the next example. □

Example 6.3.1. In the PN space of Example 2.2.1, which is not a Šerstnev PN space, since the set $\{\frac{1}{n} : n \in \mathbb{N}\}$ has a strong Cauchy subsequence, it is probabilistic strongly totally bounded but it is not a $\mathcal{D}$-bounded set (note that $\nu_p(x) = exp(-\sqrt{|p|}) < 1$, for all $p \neq 0$ and that in this space only $\{0\}$ is a $\mathcal{D}$-bounded set).

Lemma 6.3.4. *In a strict Šerstnev PN space (V, ν, τ) every strong Cauchy sequence is a topologically bounded set.*

Proof. Let (p_m) be a strong Cauchy sequence. Then there exists an n_0 such that for every $m, n \geq n_0$, $\nu_{p_m - p_n} \geq \varepsilon_{\frac{1}{m+n}}$. Now let $\alpha_m \to 0$ and $0 < \alpha_m < 1$, then one has

$$\begin{aligned}\nu_{\alpha_m p_m} &\geq \tau(\nu_{\alpha_m(p_m - p_{n_0})}, \nu_{\alpha_m p_{n_0}}) \geq \tau(\nu_{(p_m - p_{n_0})}, \nu_{\alpha_m p_{n_0}}) \\ &\geq \tau(\varepsilon_{\frac{1}{m+n_0}}, \nu_{\alpha_m p_{n_0}}) \to \tau(\varepsilon_0, \varepsilon_0) = \varepsilon_0,\end{aligned}$$

as m tends to infinity. □

Lemma 6.3.5. *In a strict Šerstnev PN space (V, ν, τ) every probabilistic strongly totally bounded set is $\mathcal{D}$-bounded.*

Proof. We show by *reductio ad absurdum* that if A is a probabilistic strongly totally bounded set then it is topologically bounded, and so by Example 6.2.7 and Theorem 6.2.11, it is $\mathcal{D}$-bounded.

If A is not topologically bounded, there exists a sequence $(p_m) \subseteq A$ and a real sequence $\alpha_m \to 0$ such that $(\alpha_m p_m)$ does not tend to the null vector in V. There is an infinite set $J \subseteq \mathbb{N}$ such that the sequence $(\alpha_m p_m)_{m \in J}$ stays off a neighborhood of the origin. Since (p_m) is probabilistic strongly totally bounded then has a Cauchy subsequence, say (p_{m_l}) which by Lemma 6.3.4 is topologically bounded, and since $\alpha_{m_l} \to 0$ then $\nu_{\alpha_{m_l} p_{m_l}} \to \varepsilon_0$ and hence $(\alpha_{m_l} p_{m_l})$ is a strong Cauchy subsequence of $(\alpha_m p_m)$. Then $(\alpha_m p_m)$ is probabilistic strongly totally bounded and so is $(\alpha_m p_m)_{m \in J}$, therefore there is a strong Cauchy subsequence of $(\alpha_m p_m)_{m \in J}$, say $(\alpha_{m_k} p_{m_k})$ which stays off a neighborhood of the origin, hence it does not tend to the null vector in V, on the other hand, since $(\alpha_{m_k} p_{m_k})$ is a strong Cauchy sequence; then there is a $k_0 \in \mathbb{N}$ such that for every $k, t \geq k_0$ one has $\nu_{p_{m_k} - p_{m_t}} \geq \varepsilon_{\frac{1}{k+t}}$. Thus

$$\begin{aligned}\nu_{\alpha_{m_k} p_{m_k}} &\geq \tau(\nu_{\alpha_{m_k}(p_{m_k} - p_{m_{k_0}})}, \nu_{\alpha_{m_k} p_{m_{k_0}}}) \geq \tau(\nu_{p_{m_k} - p_{m_{k_0}}}, \nu_{\alpha_{m_k} p_{m_{k_0}}}) \\ &\geq \tau(\varepsilon_{\frac{1}{k+t}}, \nu_{\alpha_{m_k} p_{m_k}}) \to \tau(\varepsilon_0, \varepsilon_0) = \varepsilon_0,\end{aligned}$$

as k tends to $+\infty$, which is a contradiction. □

Every $\mathcal{D}$-bounded set is not a probabilistic strongly totally bounded set, in general, as can be seen from the next example.

Example 6.3.2. Let $\nu : l^\infty \to \Delta^+$ via $\nu_p := \varepsilon_{\|p\|}$ for every $p = (p_n)_{n \geq 0} \in l^\infty$, let τ, τ^* be continuous triangle functions such that $\tau \leq \tau^*$ and

$\tau(\varepsilon_a, \varepsilon_b) = \varepsilon_{a+b}$, for all $a, b > 0$. For instance, it suffices to take $\tau = \tau_T$ and $\tau^* = \tau_{T^*}$, where T is a continuous t-norm and T^* is its t-conorm. Then $(l^\infty, \nu, \tau, \tau^*)$ is a PN space. Suppose

$$A = \{p : \|p\|_\infty = 1, p \in l^\infty\},$$

the set A is $\mathcal{D}$-bounded but not a probabilistic strongly totally bounded set. In fact $R_A(x) = \lim_{y\to x, y<x} \inf\{\varepsilon_{\|p\|}(y) : p \in A\} \to 1$ as $x \to +\infty$. Let $(p_n)_{n\in\mathbb{N}}$ be a sequence of A, where

$$p_1 = (1, 0, 0, \ldots, 0, \ldots), \quad p_2 = (0, 1, 0, \ldots, 0, \ldots), \ldots,$$
$$p_n = (0, 0, 0, \ldots, 1, 0, \ldots), \ldots$$

in view of the definition of a strong Cauchy sequence, it is obvious that $(p_n)_{n\in\mathbb{N}}$ is not a strong Cauchy sequence. And by Theorem 6.3.1, one has that A is not a probabilistic strong totally bounded set.

Theorem 6.3.2. *Let (V, ν, τ, τ^*) be a PN space. If A and B are two probabilistic strongly totally bounded subsets of V, then the following statements hold:*

(i) *$A \cup B$ is a probabilistic strongly totally bounded subset of V;*
(ii) *$A + B$ is a probabilistic strongly totally bounded subset of V, where the set $A + B$ is defined via $A + B := \{p + q : p \in A, q \in B\}$.*

Proof.

(i) By Definition 6.3.1 for every $F \in \mathcal{D}^+$ and $F < \varepsilon_0$, there exists a finite subset S_F of A and S'_F of B such that $A \subseteq \bigcup_{p\in S_F} D_p(F)$ and $B \subseteq \bigcup_{p\in S'_F} D_p(F)$, where $D_p(F) = \{q \in V : \nu_{p-q} > F\}$. So we have that $A \cup B \subseteq \bigcup_{p\in S_F} D_p(F) \bigcup (\bigcup_{p\in S'_F} D_p(F)) = \bigcup_{p\in S_F \cup S'_F} D_p(F)$, which is the assertion.

(ii) Let $(c_n)_{n\in\mathbb{N}}$ be a sequence of $A + B$. Suppose $c_n = p_n + q_n$, where $(p_n) \subset A$ and $(q_n) \subset B$. Because A and B are probabilistic strongly totally bounded subsets, by Theorem 6.3.1 there exists a subsequence $p_{k,n}$ of (p_n) and $q_{k,n}$ of (q_n), such that $p_{k,n}$ and $q_{k,n}$ are both strong Cauchy subsequences, i.e. $\lim_{m,n\to\infty} \nu_{p_{k,n}-p_{k,m}} = \varepsilon_0$ and $\lim_{m,n\to\infty} \nu_{q_{k,n}-q_{k,m}} = \varepsilon_0$, so that

$$\begin{aligned}\nu_{c_{k,n}-c_{k,m}} &= \nu_{(p_{k,n}+q_{k,n})-(p_{k,m}+q_{k,m})} = \nu_{(p_{k,n}-p_{k,m})+(q_{k,n}-q_{k,m})} \\ &\geq \tau(\nu_{p_{k,n}-p_{k,m}}, \nu_{p_{k,n}-p_{k,m}}) \to \tau(\varepsilon_0, \varepsilon_0) = \varepsilon_0,\end{aligned}$$

as m, n tends to infinity, i.e., the subsequence $(c_{n,k})$ of (c_n) is a strong Cauchy subsequence. Finally, by Theorem 6.3.1 one has that $A + B$ is a probabilistic strongly totally bounded subset of V. □

Corollary 6.3.1. *Let (V,ν,τ,τ^*) be a PN space. Let $\{A_i\}_{i=1}^{i=n}$ be a family of a finite number of probabilistic strongly totally bounded subsets of V. Define $\sum_{i=1}^n A_i$ via $\sum_{i=1}^n A_i := A_1 + A_2 + \cdots + A_n$. Then one has that $\cup_{i=1}^n A_i$ and $\sum_{i=1}^n A_i$ are both probabilistic strongly totally bounded subsets of V.*

6.4 $\mathcal{D}$-Compact Sets in PN Spaces

Lemma 6.4.1. *A $\mathcal{D}$-compact subset of a PN space (V,ν,τ,τ^*) in which $\nu(V) \subseteq \mathcal{D}^+$ and $\mathcal{D}^+$ is stable under τ, is $\mathcal{D}$-bounded and closed.*

Proof. Suppose that A is $\mathcal{D}$-compact. If A is $\mathcal{D}$-unbounded it contains a $\mathcal{D}$-unbounded sequence $(p_m)_{m\in\mathbb{N}}$ such that $\nu_{p_m} < \varepsilon_m$. This sequence could not have a convergent subsequence, since a convergent sequence must be $\mathcal{D}$-bounded by Theorem 6.2.11. The closedness of A is trivial. □

Lemma 6.4.2. *Let (V,ν,τ,τ^*) be a TV PN space. If A is a $\mathcal{D}$-compact subset of V, then A is topologically bounded.*

Proof. Suppose that A is a $\mathcal{D}$-compact subset of (V,ν,τ,τ^*). Let (p_n) be an arbitrary sequence of elements of A and (a_n) an arbitrary sequence of real numbers that converges to 0, without loss of generality in assuming $0 \le |a_n| \le 1$ for every $n \in \mathbf{N}$. Since A is a $\mathcal{D}$-compact subset of V, from Definition 3 (iii) we know that (p_n) has a convergent subsequence (p_{n_k}) which is convergent to an element $p \in A$. Without loss of generality, we suppose that the convergent subsequence of (p_n) is itself. From Lemma 3.3.1, Theorem 6.2.8, and Definition 2.2.1 (N3), one has that

$$\begin{aligned}\nu_{a_n p_n} &\ge \tau(\nu_{a_n(p_n-p)}, \nu_{a_n p}) \ge \tau(\nu_{(p_n-p)}, \nu_{a_n p})\\ &\to \tau(\varepsilon_0, \varepsilon_0) = \varepsilon_0,\end{aligned}$$

as n tends to ∞, i.e., $\nu_{a_n p_n}(x) \to 1$ as $n \to +\infty$ for all $x > 0$. Thus $a_n p_n \to 0$ in the strong topology, from which one concludes that A is topologically bounded. □

As in the classical case, a $\mathcal{D}$-bounded and closed subset of a (finite dimensional) PN space is not $\mathcal{D}$-compact in general, as one can see from the next examples.

Example 6.4.1. We consider a quadruple $(\mathbb{Q},\nu,\tau_\Pi,\tau_T)$, where $\prod(a,b) = a\cdot b$, $T(a,b) = \frac{1}{1+[(1/a-1)^2+(1/b-1)^2]^{1/2}}$, for every $a,b \in (0,1)$ and the probabilistic norm is defined via $\nu_p(t) := \frac{t}{t+|p|^{1/2}}$. It is straightforward to check that $(\mathbb{Q}, \nu, \tau_{\prod}, \tau_T)$ is a PN space. In this space, convergence of a sequence is equivalent to its convergence in $\mathbb{R}$. One considers the subset $A = [a, b] \cap \mathbb{Q}$, where $a, b \in \mathbb{R}\backslash\mathbb{Q}$. Since $R_A(t) = \frac{t}{t+(\max\{|a|,|b|\})^{1/2}}$, then the subset A is $\mathcal{D}$-bounded and since A is closed in $\mathbb{Q}$ classically, so it is closed in $(\mathbb{Q}, \nu, \tau_{\prod}, \tau_T)$. We know A is not classically compact in $\mathbb{Q}$, i.e., there exists a sequence in $\mathbb{Q}$ with no convergent subsequence in a classical sense and so in $(\mathbb{Q}, \nu, \tau_{\prod}, \tau_T)$. Hence A is not $\mathcal{D}$-compact.

Example 6.4.2. One considers the PN space $(\mathbb{R}, \nu, \tau, \prod_M)$ introduced in Example 3.3.3, with $\tau(\varepsilon_c, \varepsilon_d) \leq \varepsilon_{c+d}$, $(c > 0, d > 0)$, its probabilistic norm being $\nu_p := \varepsilon_{\frac{|p|}{a+|p|}}$ for every $p \in \mathbb{R}$ and for a fixed $a > 0$. With this probabilistic norm, $\mathbb{R}$ is $\mathcal{D}$-bounded and closed. But $\mathbb{R}$ is not $\mathcal{D}$-compact, because the sequence $(2^m)_{m\in\mathbb{N}}$ in $\mathbb{R}$ does not have any convergent subsequence in this space.

Example 6.4.3. The PN space $(\mathbb{R}, \nu, \tau, \prod_M)$ where the probabilistic norm is a map $\nu : \mathbb{R} \to \Delta^+$ such that $\nu_0 = \varepsilon_0$, $\nu_p := \varepsilon_{\frac{a+|p|}{a}}$ if $p \neq 0$, $(a > 0)$, and $\tau(\varepsilon_c, \varepsilon_d) \leq \varepsilon_{c+d}$, $(c > 0, d > 0)$, is a PN space (see Example 3.3.2). If A is a non-empty, classically bounded set in $\mathbb{R}$, then there exists $s > 0$ such that for every $p \in A$ one has $|p| \leq s$, since $\nu_p \geq \varepsilon_{\frac{a+s}{a}}$, A is $\mathcal{D}$-bounded. It also is trivial that A is closed. Now we show that A is not $\mathcal{D}$-compact: assume, if possible, that A is $\mathcal{D}$-compact and $(p_m)_{m\in\mathbb{N}}$ is an arbitrary sequence in A which has a subsequence (p_{m_k}) convergent to some $p \in A$, then one has

$$\lim_{k\to+\infty} \nu_{p_{m_k}-p} = \lim_{k\to+\infty} \varepsilon_{\frac{a+|p_{m_k}-p|}{a}} \neq \varepsilon_0.$$

This implies that p is not in A, which is a contradiction.

6.5 Finite Dimensional PN Spaces

In this section, we are interested in some properties of a finite dimensional PN space; in particular we introduce the definition of equivalent norms in a PN space.

Theorem 6.5.1. *Let $\{p_1, \ldots, p_n\}$ be a linearly independent set of vectors in a PN space (V, ν, τ, τ^*) such that τ^* is Archimedean and $\nu_p \neq \varepsilon_\infty$ for every $p \in V$. Then there is a number $c \neq 0$ and there exists a probabilistic norm $\nu' : \mathbb{R} \to \Delta^+$ on the real PN space $(\mathbb{R}, \nu', \tau', \tau'^*)$ where τ'^* is Archimedean and $\nu'_p \neq \varepsilon_\infty$, such that for every choice of n real scalars $\alpha_1, \cdots, \alpha_n$ one has*

$$\nu_{\alpha_1 p_1+\cdots+\alpha_n p_n} \leq \nu'_{c(|\alpha_1|+\cdots+|\alpha_n|)}. \tag{6.5.1}$$

Proof. We write $s = |\alpha_1| + \cdots + |\alpha_n|$. If $s = 0$, all α_j $(j = 1, \cdots, n)$ are zero, so (6.5.1) holds. Assume $s > 0$. One defines $\mu_p = \nu_{sp}$ and $\mu'_r = \nu'_{sr}$. Then (6.5.1) is equivalent to the following inequality,

$$\mu_{\beta_1 p_1 + \cdots + \beta_n p_n} \leq \mu'_c, \quad \beta_j = \frac{\alpha_j}{s}, \quad \left(\sum_{j=1}^{n} |\beta_j| = 1 \right). \tag{6.5.2}$$

Hence it suffices to prove the existence of $c \neq 0$ and μ' such that (6.5.2) holds. Suppose otherwise, then there exists a sequence (q_m) of vectors

$$q_m = \beta_1^{(m)} p_1 + \cdots + \beta_n^{(m)} p_n, \quad \left(\sum_{j=1}^{n} |\beta_j^{(m)}| = 1 \right),$$

such that $\mu_{q_m} \to \varepsilon_0$ as $m \to \infty$. Since $(\Sigma_{j=1}^{n} |\beta_j^{(m)}| = 1)$, one has $|\beta_j^{(m)}| \leq 1$. Hence, the sequence $(\beta_1^{(m)})$ has a convergent subsequence. Let β_1 denote the limit of such a subsequence, and let $(q_{1,m})$ denote the corresponding subsequence of (q_m). By the same argument, $(q_{1,m})$ has a subsequence $(q_{2,m})$ for which the corresponding sequence of real scalars $(\beta_2^{(m)})$ converges say to β_2. Continuing this process, we obtain a subsequence $(q_{n,m})$ of (q_m) such that

$$q_{n,m} = \sum_{j=1}^{n} \gamma_j^{(m)} p_j, \quad \left(\sum_{j=1}^{n} |\gamma_j^{(m)}| = 1 \right),$$

and $\gamma_j^{(m)} \to \beta_j$ as $m \to \infty$. Hence

$$\lim_{m \to \infty} q_{n,m} = q := \sum_{j=1}^{n} \beta_j p_j,$$

where $\sum_{j=1}^{n} |\beta_j| = 1$, since

$$\begin{aligned} \mu_{q_{n,m} - q} &= \mu_{\Sigma_{j=1}^{n} (\gamma_j^{(m)} - \beta_j) p_j} \\ &\geq \tau^{n-1} (\mu_{(\gamma_1^{(m)} - \beta_1) p_1}, \ldots, \mu_{(\gamma_n^{(m)} - \beta_n) p_n}) \to \varepsilon_0 \end{aligned}$$

as m tends to ∞. Since $\{p_1, \ldots, p_n\}$ is a linearly independent set of vectors and not all $\beta'_j s$ are zero, we have $q \neq 0$. Since $\lim_{m \to \infty} \mu_{q_m} = \varepsilon_0$, one has $\mu_{q_{n,m}} \to \varepsilon_0$. But

$$\mu_q = \mu_{(q - q_{n,m}) + q_{n,m}} \geq \tau(\mu_{q - q_{n,m}}, \mu_{q_{n,m}}) \to \varepsilon_0,$$

and hence $q = 0$, which contradicts $q \neq 0$. □

The following example shows that in the above theorem we need the field $\mathbb{R}$ to be a strong TV space.

Example 6.5.1. Consider the PN space $(\mathbb{R}, \nu, \tau, \tau^*)$ where τ^* is Archimedean and $\nu_p \neq \varepsilon_\infty$. By the above theorem there exists a $c \neq 0$ and a probabilistic norm $\nu' : \mathbb{R} \to \Delta^+$ such that $\nu_p \leq \nu'_{cp}$. If in the PN space $(\mathbb{R}, \nu', \tau', \tau'^*)$ $\lim_m \nu'_{p_m} < \varepsilon_0$ whenever $p_m \to 0$ in $\mathbb{R}$, then for the sequence (2^{-n}) one has $\nu_{2^{-n}} \leq \nu'_{c2^{-n}}$ and consequently $\varepsilon_0 < \varepsilon_0$, which is a contradiction.

From now on all the fields are strong TV spaces.

Theorem 6.5.2. *Every finite dimensional subspace W of a PN space (V, ν, τ, τ^*) where τ^* is Archimedean and $\nu_p \neq \varepsilon_\infty$ for every $p \in V$, is complete. In particular, every finite dimensional PN space is complete.*

Proof. Let (q_m) be a strong Cauchy sequence in W with $\dim W = n$; and let $\{w_1, \ldots w_n\}$ be a linearly independent subset of W. Then each q_m has a unique representation of the form

$$q_m = \alpha_1^{(m)} w_1 + \cdots + \alpha_n^{(m)} w_n.$$

Since (q_m) is a strong Cauchy sequence, for every $h > 0$ there is a positive integer N such that

$$\nu_{q_m - q_k}(h) > 1 - h,$$

whenever $m, k \geq N$. By the above theorem and Lemma 3.3.1, one has, for every $j = 1, \ldots, n$,

$$\begin{aligned} 1 - h < \nu_{q_m - q_k}(h) &= \nu_{\Sigma_{j=1}^n (\alpha_j^m - \alpha_j^k) w_j}(h) \\ &\leq \nu'_{c\Sigma_{j=1}^n |\alpha_j^m - \alpha_j^k|}(h) \leq \nu'_{c|\alpha_j^m - \alpha_j^k|}(h), \end{aligned}$$

where $c \neq 0$, $\nu' : \mathbb{R} \to \Delta^+$ and $m, k \geq N$. This shows that each of the n sequences $(\alpha_j^{(m)})_m$ where $j = 1, \cdots, n$ is a strong Cauchy in $(\mathbb{R}, \nu, \tau, \tau^*)$. Hence it converges, say to α_j. Now let us define $q = \alpha_1^{(m)} w_1 + \cdots + \alpha_n^{(m)} w_n$. Clearly $q \in W$. Furthermore,

$$\begin{aligned} \nu_{q_m - q} &= \nu_{\Sigma_{j=1}^n (\alpha_j^{(m)} - \alpha_j) w_j} \\ &\geq \tau^{n-1}(\nu_{(\alpha_1^{(m)} - \alpha_1) w_1}, \cdots, \nu_{(\alpha_n^{(m)} - \alpha_j) w_n}), \end{aligned}$$

that tends to ε_0 whenever $m \to \infty$. This means that q_m converges strongly to q. Hence W is complete. □

In the following we introduce the definition of equivalent norms in a PN space and prove that in a finite dimensional probabilistic normed space, every two probabilistic norms are equivalent.

Definition 6.5.1. A probabilistic norm $\nu : V \to \Delta^+$ is said to be equivalent to a probabilistic norm $\mu : V \to \Delta^+$, if for every sequence $(p_m)_{m\in\mathbb{N}}$ in V, $p_m \xrightarrow{\nu} p$ if, and only if, $p_m \xrightarrow{\mu} p$

In the following example two equivalent norms in PN spaces are given.

Example 6.5.2. Let us consider two PN spaces, (V, ν, τ, τ^*) with probabilistic norm $\nu_p = \varepsilon_{\|p\|}$ where $\tau(\varepsilon_c, \varepsilon_d) = \varepsilon_{c+d}, (c > 0, d > 0)$ and (V, μ, τ', Π_M) with probabilistic norm $\mu_p = \varepsilon_{\frac{\|p\|}{a+\|p\|}}, a > 0$ (see Example 3.3.3). These two probabilistic norms are equivalent and so identity map $J : V \to V$ is continuous.

Theorem 6.5.3. *Let (V, ν, τ, τ^*) and (V, μ, τ', τ'^*) be two finite dimensional PN spaces. Then, every two probabilistic norms, ν of (V, ν, τ, τ^*) and μ of (V, μ, τ', τ'^*) are equivalent whenever τ^* and τ'^* are Archimedean, $\nu_p \neq \varepsilon_\infty$ and $\mu_p \neq \varepsilon_\infty$, for every $p \in V$.*

Proof. Let $\{w_1, \ldots, w_n\}$ be a linearly independent subset of V, and consider $p_m \xrightarrow{\nu} p$. We know that both p_m and p have a unique representation as

$$p_m = \alpha_1^{(m)} w_1 + \cdots + \alpha_n^{(m)} w_n \quad \text{and} \quad p = \alpha_1 w_1 + \cdots + \alpha_n w_n.$$

By Theorem 6.5.1 and Lemma 3.3.1, one has

$$\nu_{p_m - p} = \nu_{\sum_{j=1}^n (\alpha_j^{(m)} - \alpha_j) w_j} \leq \nu'_{c \sum_{j=1}^n |\alpha_j^{(m)} - \alpha_j|} \leq \nu'_{c|\alpha_j^{(m)} - \alpha_j|},$$

where $c \neq 0$ and $\nu' : \mathbb{R} \to \Delta^+$. Therefore $\lim_m \nu'_{c(|\alpha_j^{(m)} - \alpha_j|)} = \nu'_{c(\lim_m |\alpha_j^{(m)} - \alpha_j|)} = \varepsilon_0$, that is $\alpha_j^{(m)} \to \alpha_j$ in $\mathbb{R}$. But

$$\mu_{p_m - p} = \mu_{\sum_{j=1}^n (\alpha_j^{(m)} - \alpha_j) w_j} \geq \tau'^{n-1}(\mu_{(\alpha_1^{(m)} - \alpha_1) w_1}, \ldots, \mu_{(\alpha_n^{(m)} - \alpha_1) w_n}),$$

so by continuity of τ' one has $p_m \xrightarrow{\mu} p$. By the same argument $p_m \xrightarrow{\mu} p$ implies $p_m \xrightarrow{\nu} p$. □

In the next example we show that there are two PN spaces which are not equivalent even in a finite dimensional PN space. Indeed, since τ_M is not Archimedean, the first PN space in the next example is not a strong TV space.

Example 6.5.3. Consider the PN space $(\mathbb{R}, \nu, \tau_W, \tau_M)$ where $\nu_0 = \varepsilon_0$ and $\nu_p = \frac{1}{|p|+2}\varepsilon_0 + \frac{|p|+1}{|p|+2}\varepsilon_\infty$ for $p \neq 0$: we know τ_M is not Archimedean, (see Example 3.3.1 taken from Alsina, Schweizer, and Sklar (1997)) and the PN space $(\mathbb{R}, \nu, \tau, \tau^*)$ with probabilistic norm $\nu_p = \varepsilon_{|p|}$ where $\tau(\varepsilon_c, \varepsilon_d) = \varepsilon_{c+d}$, $(c > 0, d > 0)$, (this example is taken from Lafuerza-Guillén, and Rodríguez Lallena, and Sempi (1999)). Now, the sequence $(1/n)$ in the first PN space is not convergent, but in the second one it is convergent. Therefore the above PN spaces are not equivalent.

Theorem 6.5.4. *Consider a finite dimensional PN space (V, ν, τ, τ^*) where τ^* is Archimedean, $\nu_p \neq \varepsilon_\infty$, $\nu(V) \subseteq \mathcal{D}^+$, and $\mathcal{D}^+$ is stable under τ, for every $p \in V$ and the real PN space $(\mathbb{R}, \nu', \tau', \tau'^*)$, where ν' has the LG-property, then every subset A of V is $\mathcal{D}$-compact if, and only if A is $\mathcal{D}$-bounded and closed.*

Proof. By Lemma 6.4.1 $\mathcal{D}$-compact subsets A of V are $\mathcal{D}$-bounded and closed, so it remains only to prove the converse. Let $\dim V = n$ and $\{w_1, \ldots, w_n\}$ be a linearly independent subset of V. Consider any sequence (q_m) in A. Each q_m has a representation $q_m = \alpha_1^{(m)} w_1 + \cdots + \alpha_m^{(m)} w_n$. Since A is $\mathcal{D}$-bounded so is (q_m), and so there exists a *d.f.* $G \in \mathcal{D}^+$ such that

$$G \leq \nu_{q_m} \leq \nu_{\alpha_1^{(m)} w_1 + \cdots + \alpha_m^{(m)} w_n}.$$

By Theorem 6.5.1 and Lemma 3.3.1, one has

$$\nu_{\alpha_1^{(m)} w_1 + \cdots + \alpha_m^{(m)} w_n} \leq \nu'_{c(|\alpha_1^{(m)}| + \cdots + |\alpha_n^{(m)}|)} \leq \nu'_{c|\alpha_j^{(m)}|}.$$

Hence for each fixed j, the sequence $(\alpha_j^{(m)})$ is $\mathcal{D}$-bounded and since ν' has the LG-property, then by Lemma 6.2.2, it is also classically bounded. Therefore for every $1 \leq j \leq n$ the sequence $(\alpha_j^{(m)})_m$ has a convergent subsequence that converges to some α_j, $j = 1, \cdots, n$. As in the proof of Theorem 6.5.1, we can construct a subsequence (r_m) of (q_m) which converges to $r := \sum_{j=1}^{n} \alpha_j w_j$. Since A is closed, $r \in A$. This shows that each sequence (q_m) in A has a convergent subsequence in A. Thus A is $\mathcal{D}$-compact. □

Chapter 7

Normability

The normability of PN spaces has been recently introduced, first in Lafuerza-Guillén *et al.* (2008), and soon after partially developed in Zhang and Zhang (2008).

If (V, ν, τ, τ^*) is a TV space, the question naturally arises of whether it is also normable; in other words, whether there is a norm on V that generates the strong topology. This question had been broached by Prochaska (1967) in the case of Šerstnev PN spaces.

In this chapter we look more closely at the issue of developing the topology of generalized Šerstnev spaces (henceforth, PN spaces according to Definition 2.2.1). Relying on Kolmogorov's classical characterization of normable TV spaces, we study the normability of those PN spaces that are also TV spaces and provide a complete characterization of those strict Šerstnev spaces that are indeed normable. We will give several sufficient conditions, under which many PN spaces are normable. Up to now, there exists little information about the normability of PN spaces. In the process, we shall need Kolmogorov's classical characterization of normability for T_1 spaces (Kolmogorov, 1934, 1991). As an application of our results several examples are given which are normable PN spaces, but not Šerstnev spaces.

7.1 Normability of Šerstnev Spaces

Theorem 7.1.1 (Kolmogorov). *A T_1 TV space is normable if, and only if, there is a neighborhood of the origin θ that is convex and topologically bounded.*

Readers are reminded that a space of Šerstnev (V, ν, τ) is *strict* whenever $\nu(V) \subseteq \mathcal{D}^+$. In characterizing normable Šerstnev spaces we shall need the following result.

Theorem 7.1.2. *In a strict Šerstnev space (V, ν, τ) the following statements are equivalent for a subset A of V:*

(a) *A is $\mathcal{D}$-bounded;*
(b) *A is topologically bounded.*

Proof. (a) $\Longrightarrow$ (b) Let A be any $\mathcal{D}$-bounded subset of V and let (p_n) be any sequence of elements of A and (λ_n) any sequence of real numbers that converges to 0; without loss of generality assume $\lambda_n \neq 0$ for every $n \in \mathbf{N}$. Then, for every $x > 0$, and for every $n \in \mathbf{N}$,

$$\nu_{\lambda_n p_n}(x) = \nu_{p_n}\left(\frac{x}{|\lambda_n|}\right) \geq R_A\left(\frac{x}{|\lambda_n|}\right) \xrightarrow[n \to +\infty]{} 1.$$

Thus $\lambda_n p_n \to \theta$ in the strong topology and A is topologically bounded.
(b)$\Longrightarrow$(a) Let A be a subset of V which is not $\mathcal{D}$-bounded. Then

$$R_A(x) \xrightarrow[x \to +\infty]{} \gamma < 1.$$

By definition of R_A, for every $n \in \mathbf{N}$ there is $p_n \in A$ such that

$$\nu_{p_n}(n^2) < \frac{1+\gamma}{2} < 1.$$

If $\lambda_n = 1/n$, then, for every $n \in \mathbf{N}$,

$$\nu_{\lambda_n p_n}(1/2) \leq \nu_{\lambda_n p_n}(n) = \nu_{p_n}(n^2) < \frac{1+\gamma}{2} < 1,$$

which shows that $(\nu_{\lambda_n p_n})$ does not tend to φ_0, even if it has a weak limit, *viz.* $(\lambda_n p_n)$ does not tend to θ in the strong topology; in other words, A is not topologically bounded. □

As a consequence of the previous results, it is now possible to characterize normability for strict Šerstnev spaces according to the following criterion.

Theorem 7.1.3. *A strict Šerstnev space (V, ν, τ) is normable if, and only if, the null vector θ has a convex $\mathcal{D}$-bounded neighbourhood.*

We prove here the following (restrictive) sufficient condition (see Prochaska (1967)) not only for the sake of completeness, but also because Prochaska's thesis is not easily accessible and, moreover, because the notation there adopted is different from the one that has become usual after the publication of Schweizer and Sklar's (2005) *Probabilistic Metric Spaces.*

Theorem 7.1.4 (Prochaska). *A strict Šerstnev space* (V, ν, τ) *with* $\tau = \tau_M$ *is locally convex.*

Proof. It suffices to consider the family of neighborhoods of the origin θ, $N_\theta(t)$, with $t > 0$. Let $t > 0$, $p, q \in N_\theta(t)$ and $\lambda \in [0, 1]$. Then

$$\begin{aligned}\nu_{\lambda p+(1-\lambda)q}(t) &\geq \tau_M(\nu_{\lambda p}, \nu_{(1-\lambda)q})(t) \\ &= \sup_{\mu\in[0,1]} M(\nu_{\lambda p}(\mu t), \nu_{(1-\lambda)q}((1-\mu)t)) \\ &\geq M(\nu_{\lambda p}(\lambda t), \nu_{(1-\lambda)q}((1-\lambda)t)) = M(\nu_p(t), \nu_q(t)) > 1 - t.\end{aligned}$$

Thus $\lambda p + (1 - \lambda)q$ belongs to $N_\theta(t)$ for every $\lambda \in [0, 1]$. □

As a consequence of Theorems 7.1.1 and 7.1.4, every simple PN space $(V, \|\cdot\|, G, M)$ with $G \in \mathcal{D}^+$ is trivially normable, since their strong topology coincides with the topology of their classical norm. In general, it is to be expected that most of the PN spaces considered in Theorem 7.1.4 will be normable, as shown by the following corollary.

Corollary 7.1.1. *Let* (V, ν, τ_M) *be a strict Šerstnev space. If* $N_\theta(t)$ *is* $\mathcal{D}$*-bounded for some* $t \in]0, 1[$, *then* (V, ν, τ_M) *is normable.*

7.2 Other Cases

Apart from the Šerstnev spaces, we can also determine whether an α-simple space is normable, as the following result shows.

Theorem 7.2.1. *Let* G *be a continuous and strictly increasing distribution function in* $\mathcal{D}^+$. *Then, the* α*-simple space* $(V, \|\cdot\|, G; \alpha)$ *is normable.*

Proof. It follows from the assumptions that the α-simple space $(V, \|\cdot\|, G; \alpha)$ is a Menger space under a suitable t-norm T; see Lafuerza-Guillén *et al.* (1997). Let $N_\theta(t)$ be a neighborhood of the origin θ with $t \in]0, 1[$; then

$$\begin{aligned}N_\theta(t) &= \left\{p \in V \mid G\left(\frac{t}{\|p\|^\alpha}\right) > 1 - t\right\} \\ &= \left\{p \in V \mid \|p\| < \left(\frac{t}{G^{-1}(1-t)}\right)^{1/\alpha}\right\}.\end{aligned}$$

Since $h(t) = (t/G^{-1}(1-t))^{1/\alpha}$ is a continuous function such that $\lim_{t\to 0+} h(t) = 0$ and $\lim_{t\to 1-} h(t) = \infty$, then it is clear that the strong topology for V coincides with the topology of the norm $\|\cdot\|$ in V. Therefore, $(V, \|\cdot\|, G; \alpha)$ is normable. □

It is natural to ask whether results similar to that of Theorem 7.1.3 hold for general PN spaces. The conditions of Theorem 7.1.2 need not be equivalent; for, there are PN spaces in which a set A may be topologically bounded without being $\mathcal{D}$-bounded. On the other hand, even in those cases, it is sometimes possible to establish directly whether a PN space that is also a topological vector space is normable. To illustrate both facts, we next introduce a new class of PN spaces whose interest goes deeper than just to provide an example to this point. Finding a new type of PN spaces is useful in order to deepen our knowledge of these spaces.

Before introducing a new class of PN spaces we need the following technical lemma.

Lemma 7.2.1. *Let $f\colon [0,+\infty[\to [0,1]$ be a right-continuous non-increasing function. Define $f^{[-1]}(1) := 0$ and $f^{[-1]}(y) := \sup\{x \mid f(x) > y\}$ for all $y \in [0,1[$ ($f^{[-1]}(y)$ might be infinite). For $x_0 \in [0,+\infty[$ and $y_0 \in [0,1]$, the following facts are equivalent:* (a) $f(x_0) > y_0$; (b) $x_0 < f^{[-1]}(y_0)$.

Proof. If $f(x_0) > y_0$ then $f^{[-1]}(y_0) = \sup\{x \mid f(x) > y_0\} \geq x_0$. If we suppose that $\sup\{x \mid f(x) > y_0\} = x_0$, then $f(x) \leq y_0$ for every $x > x_0$. Thus $f(x_0) = \ell^+ f(x_0) \leq y_0$ ($\ell^+ f(x_0) = \lim_{x\to x_0^+} f(x)$), against the assumption; whence (a) $\Longrightarrow$ (b). The converse result is an immediate consequence of the monotonicity of f. □

The following theorem introduces a new class of PN spaces that also provides some properties of the spaces in that class. Such properties are interesting for the purposes of this chapter. It may be useful to recall that $\tau_{T^*} \geq \tau_{M^*} = \tau_M$ for every t-norm T.

Theorem 7.2.2. *Let $(V, \|\cdot\|)$ a normed space and let T be a continuous t-norm. Let f be a function as in Lemma* 7.2.1, *and satisfying the following two properties:*

(a) $f(x) = 1$ *if, and only if,* $x = 0$;
(b) $f(\|p+q\|) \geq T(f(\|p\|), f(\|q\|))$ *for all* $p, q \in V$.

If $\nu\colon V \to \Delta^+$ *is given by*

$$\nu_p(x) = \begin{cases} 0, & x \le 0, \\ f(\|p\|), & x \in]0, +\infty[, \\ 1, & x = +\infty, \end{cases} \tag{7.2.1}$$

for every $p \in V$, *then* (V, ν, τ_T, τ_M) *is a Menger PN space satisfying the following properties:*

(F1) (V, ν, τ_T, τ_M) *is a topological vector space;*
(F2) (V, ν, τ_T, τ_M) *is normable;*
(F3) *If* $p \in V$ *and* $t > 0$, *then the strong neighborhood* $N_p(t)$ *in* (V, ν, τ_T, τ_M) *is not* $\mathcal{D}$*-bounded, but* $N_p(t)$ *is topologically bounded whenever* $N_p(t) \neq V$;
(F4) (V, ν, τ_T, τ_M) *is not a Šerstnev space;*
(F5) (V, ν, τ_T, τ_M) *is not a strict PN space.*

Proof. First, we prove that (V, ν, τ_T, τ_M) is a Menger PN space:

(N1) $\nu_p = \varphi_0 \iff f(\|p\|) = 1 \iff \|p\| = 0 \iff p = \theta$.
(N2) Trivial.
(N3) For all $p, q \in V$, the inequality $\nu_{p+q} \ge \tau_T(\nu_p, \nu_q)$ means that one has

$$\nu_{p+q}(x) \ge \tau_T(\nu_p, \nu_q)(x) = \sup_{s+t=x} T(\nu_p(s), \nu_q(t))$$

for all $x \in]0, +\infty[$, or, equivalently,

$$f(\|p+q\|) \ge T(f(\|p\|), f(\|q\|)),$$

as assumed.

(N4) Let $p \in V$ and let $\lambda \in [0, 1]$. Then, the inequality $\nu_p \le \tau_M(\nu_{\lambda p}, \nu_{(1-\lambda)p})$ is equivalent, for all $x \in]0, +\infty[$, to

$$\begin{aligned} f(\|p\|) = \nu_p(x) &\le \tau_M(\nu_{\lambda p}, \nu_{(1-\lambda)p})(x) = \sup_{s+t=x} M(\nu_{\lambda p}(s), \nu_{(1-\lambda)p}(t)) \\ &= \sup_{s+t=x} M(f(\lambda\|p\|), f((1-\lambda)\|p\|)) \\ &= M(f(\lambda\|p\|), f((1-\lambda)\|p\|)) = \min\{f(\lambda\|p\|), f((1-\lambda)\|p\|)\}. \end{aligned}$$

Therefore, one has, for all $p \in V$ and for all $\lambda \in [0, 1]$, $\nu_p \le \tau_M(\nu_{\lambda p}, \nu_{(1-\lambda)p})$ if, and only if, $f(\|p\|) \le f(\alpha\|p\|)$ for all $\alpha \in [0, 1]$, namely if, and only if, f is non-increasing.

Now we prove properties (F1) through (F5):

(F1) Let $p \in V$. We only have to prove that the map from $\mathbb{R}$ into V defined by $\lambda \mapsto \lambda p$ is continuous at every $\lambda \in \mathbb{R}$. Let $\eta > 0$ (we shall suppose, without loss of generality, that $\eta \le 1$). We must prove that there exists a number $\delta > 0$ such that $d_S(\nu_{\lambda' p - \lambda p}, \varphi_0) < \eta$ whenever $|\lambda' - \lambda| < \delta$; or, equivalently, such that $d_S(\nu_{\beta p}, \varphi_0) < \gamma$ whenever $|\beta| < \delta$. Since $d_S(\nu_q, \varphi_0) = \inf\{h \mid \ell^+ \nu_q(h) > 1 - h\} = 1 - f(\|q\|)$, then one has $d_S(\nu_{\beta p}, \varphi_0) < \gamma$ if, and only if, $1 - f(|\beta| \|p\|) < \gamma$, *viz.* $f(|\beta|\|p\|) > 1 - \gamma$, or, again, by Lemma 7.2.1, if, and only if, $|\beta| < \delta := f^{[-1]}(1 - \gamma)/\|p\|$.

(F2) Let $p \in V$. Let $t > 0$ (we shall suppose, without loss of generality, that $t < 1 - \lim_{x\to\infty} f(x)$). Then, because of Lemma 7.2.1,

$$\begin{aligned} N_p(t) &= \{q \in V \mid d_S(\nu_{p-q}, \varphi_0) < t\} = \{q \in V \mid f(\|p - q\|) > 1 - t\} \\ &= \{q \in V \mid \|p - q\| < f^{[-1]}(1 - t)\} = B(p, f^{[-1]}(1 - t)), \end{aligned}$$

i.e., the strong neighborhood $N_p(t)$ is a ball in $(V, \|\cdot\|)$ with centre at p. Conversely, let $r > 0$. If $t = 1 - f(r)$, then $f^{[-1]}(1 - t) < r$, whence

$$N_p(t) = B(p, f^{[-1]}(1 - t)) \subset B(p, r).$$

Therefore, the strong topology for $(V, \nu, \tau_T, \tau_{T^*})$ coincides with the topology of the norm in $(V, \|\cdot\|)$.

(F3) If $p \in V$ and $0 < t < 1 - \lim_{x\to\infty} f(x)$, then $N_p(t) = B(p, f^{[-1]}(1 - t))$ is a ball in $(V, \|\cdot\|)$, whence $N_p(t)$ is topologically bounded. On the other hand, if $0 < x < \infty$ then

$$\begin{aligned} \phi_{N_p(t)}(x) &= \inf\{\nu_q(x) \mid q \in N_p(t)\} \\ &= \inf\{f(\|q\|) \mid \|p - q\| < f^{[-1]}(1 - t)\} \\ &= f(\|p\| + f^{[-1]}(1 - t)). \end{aligned}$$

Thus, $\lim_{x\to\infty} R_{N_p(t)}(x) = f(\|p\| + f^{[-1]}(1 - t)) < 1$, i.e., $N_p(t)$ is not $\mathcal{D}$-bounded.

(F4) It is immediately clear that $(V, \nu, \tau_T, \tau_{T^*})$ is a Šerstnev space if, and only if, the function f is constant on $]0, \infty[$. From assumption (a) this constant should be less than 1, which contradicts the right-continuity of f at $x = 0$. Thus, (V, ν, τ_T, τ_M) is not a Šerstnev space.

(F5) It is immediate that $\nu(V \backslash \{\theta\}) \subset \Delta^+ \backslash \mathcal{D}^+$. □

Now we consider some special cases and use the preceding theorem in order to give some examples.

Example 7.2.1. Suppose that, in Theorem 6.2.2, $T = \Pi$. Then, property (b) reads $f(\|p+q\|) \geq f(\|p\|)f(\|q\|)$ for all $p, q \in V$. It is not difficult to prove that, under the given assumptions on f, property (b) is equivalent to the following:

$$f(x+y) \geq f(x)f(y), \quad \text{for all } x, y \in [0, \infty[. \tag{7.2.2}$$

The following are examples of functions f satisfying the assumptions of Theorem 6.2.2 in this case:

$$f_{\alpha,\beta}(x) := 1 - \frac{\beta}{\alpha} + \frac{\beta}{x+\alpha}, \quad 0 \leq \beta \leq \alpha,$$

$$g_{\alpha,\beta}(x) := 1 - \alpha + \alpha \exp\left(-x^{\beta}\right), \quad 0 < \alpha \leq 1, \ \beta > 0.$$

Example 7.2.2. Take $T = W$ in Theorem 6.2.2. In this case property (b) reads

$$\forall p, q \in V \quad f(\|p+q\|) \geq f(\|p\|) + f(\|q\|) - 1,$$

which is equivalent to the following one

$$\forall x, y \in [0, +\infty[\quad 1 + f(x+y) \geq f(x) + f(y),$$

namely to the fact that the function $x \mapsto f(x) - 1$ is superadditive. For instance, the following functions satisfy these properties but not those considered in Example 6.2.1, since they do not satisfy (6.2.2):

$$h_{\alpha,\beta}(x) := \begin{cases} 1 - \alpha x, & 0 \leq x \leq \beta, \\ 1 - \alpha\beta, & x > \beta, \end{cases} \quad 0 < \beta \leq 1/\alpha.$$

7.3 Normability of PN Spaces

Definition 7.3.1. Let (V, ν, τ, τ^*) be a PN space. For $p \in V$ and $\lambda \in]0,1[$ assume the following two conditions:

(Z1) For all $a \in]0,1[$, there exists a $\beta \in [1, \infty[$ such that

$$\nu_p(\lambda) > 1 - \lambda \Rightarrow \nu_{ap}(a\lambda) > 1 - \frac{a}{\beta}\lambda,$$

(Z2) For all $a \in]0,1[$, let $\beta_0(a,\lambda) = \frac{1+\sqrt{1-4a(1-a)\lambda}}{2}$, then

$$\nu_p(\lambda) > 1 - \lambda \Rightarrow \nu_{ap}(a\lambda) > 1 - \frac{a}{\beta_0(a,\lambda)}\lambda.$$

Remark 7.3.1. The points which satisfy (Z1) or (Z2), must belong to $N_p(\lambda)$. If every point of $N_p(\lambda)$ satisfies either (Z1) or (Z2), then $N_p(\lambda)$ is said to satisfy (Z1) or (Z2).

For $\lambda, a \in]0,1[$ it is not difficult to verify that

$$\beta_0(a,\lambda) = \beta_0(1-a,\lambda) \quad \text{and} \quad \beta_0(a,\lambda) \in]0,1[.$$

So

$$1 - \frac{a}{\beta}\lambda \geq 1 - a\lambda > 1 - \frac{a}{\beta_0(a,\lambda)}\lambda.$$

Thus (Z1) implies (Z2), but the converse fails as shown in the following example.

Example 7.3.1. Let $\nu : \mathbb{R} \to \Delta^+$ be defined by $\nu_0 = \varepsilon_0$ and

$$\nu_p(x) = \begin{cases} 0, & x \leq 0 \\ exp(-\sqrt{\|p\|}), & 0 < x < +\infty \\ 1, & x = +\infty). \end{cases}$$

A straightforward calculation shows that $(\mathbb{R}, \nu, \tau_\Pi, \tau_{\Pi^*})$ is a PN space, but the condition (Š) fails. We are going to prove that there exists at least one point $p_0 \in \mathbb{R}$ satisfying (Z2), but not (Z1).

Let $\lambda_0 \in]0,1[$ be given, then for all $a \in]0,1[$ there exists a $p_0 \in \mathbb{R}$ satisfying

$$\nu_{p_0}(\lambda_0) > 1 - \lambda_0 \quad \text{and} \quad \nu_{ap_0}(a\lambda_0) = 1 - a\lambda_0, \tag{7.3.1}$$

i.e.

$$\sqrt{\|p_0\|} < -\ln(1-\lambda_0) \quad \text{and} \quad \sqrt{\|p_0\|} = -\ln(1-a\lambda_0)^{\frac{1}{\sqrt{a}}}.$$

Thus it is sufficient to consider

$$(1-\lambda_0)^{\sqrt{a}} < 1 - a\lambda_0.$$

Let $\phi(a) = (1-\lambda_0)^{\sqrt{a}} - (1-a\lambda_0)$. From

$$\phi'(a) = \frac{1}{2\sqrt{a}} \ln(1-\lambda_0)(1-\lambda_0)^{\sqrt{a}} + \lambda_0,$$

one has

$$\begin{aligned}\phi''(a) &= \frac{1}{4a}[\ln(1-\lambda_0)]^2(1-\lambda_0)^{\sqrt{a}} - \frac{1}{4a\sqrt{a}} \ln(1-\lambda_0)(1-\lambda_0)^{\sqrt{a}} \\ &= -\frac{1}{4a} \ln(1-\lambda_0)(1-\lambda_0)^{\sqrt{a}} \left[\frac{1}{\sqrt{a}} - \ln(1-\lambda_0)\right] > 0.\end{aligned}$$

Thus $\varphi(a)$ is a concave function for $a \in]0,1[$. On the other hand $\phi(0) = \phi(1) = 0$. Therefore for $a \in]0,1[$, one has $\phi(a) < 0$, i.e., for all $a \in]0,1[$,

$$(1-\lambda_0)^{\sqrt{a}} < 1 - a\lambda_0,$$

and hence (7.3.1) holds. In view of (7.3.1) and Definition 7.3.1, we conclude that the point p_0 does not satisfy (Z1). But by Remark 7.3.1, one has that

$$\nu_{ap_0}(a\lambda_0) = 1 - a\lambda_0 > \lambda 1 - \frac{a}{\beta_0(a,\lambda_0)}\lambda_0,$$

i.e., p_0 satisfies (Z2).

Example 7.3.2. Let $(\mathbb{R}, \nu, \tau_\Pi, \tau_{\Pi^*})$ be the PN space of Example 7.3.1, then every strong λ-neighborhood $N_\theta(\lambda)$ of $(\mathbb{R}, \nu, \tau_\Pi, \tau_{\Pi^*})$ is topologically bounded, where $\lambda \in]0,1[$.

In fact, let $\lambda \in]0,1[$ be given. Let $\overline{N_\theta(\lambda)} = \{q \in \mathbb{R}; \nu_q(\lambda) \geq 1-\lambda\}$ and suppose the (p_n) is a sequence of $\overline{N_\theta(\lambda)}$, then $\nu_{p_n}(\lambda) = exp(-\sqrt{\|p_n\|}) \geq 1-\lambda$. So one has that $\|p_n\| \leq (-\ln(1-\lambda))^2$. Thus (p_n) has a classical convergent subsequence (p_{n_k}) in $\mathbb{R}$. Therefore one has (p_n) is convergent to p in $\mathbb{R}$, i.e., $p_{n_k} - p \to 0$ with $n_k \to \infty$ and $\|p\| \leq (-\ln(1-\lambda))^2)$. Hence one has that $\nu_p(\lambda) \geq 1-\lambda$ and for any $x > 0$,

$$\nu_{p_{n_k}-p}(x) = exp(-\sqrt{\|p_{n_k} - p\|}) \to 1,$$

with $n_k \to +\infty$, i.e. the sequence (p_n) in $\overline{N_\theta(\lambda)}$ has a convergent subsequence (p_{n_k}) to an element $p \in \overline{N_\theta(\lambda)}$. From Definition 3.1.1 (iii) we know that $\overline{N_\theta(\lambda)}$ is $\mathcal{D}$-compact in $(\mathbb{R}, \nu, \tau_\Pi, \tau_{\Pi^*})$.

Because of τ_{Π^*} is Archimedean, we know that $(\mathbb{R}, \nu, \tau_\Pi, \tau_{\Pi^*})$ is a TV PN space. Thus from Lemma 6.4.2, $\overline{N_\theta(\lambda)}$ with $\lambda \in]0,1[$ is topologically bounded in $(\mathbb{R}, \nu, \tau_\Pi, \tau_{\Pi^*})$. Since $N_\theta(\lambda) \subseteq \overline{N_\theta(\lambda)}$, by Remark 3.1, $N_\theta(\lambda)$ with $\lambda \in]0,1[$ is topologically bounded in $(\mathbb{R}, \nu, \tau_\Pi, \tau_{\Pi^*})$.

Theorem 7.3.1. *Let (V, ν, τ, τ^*) be a PN space and $N_\theta(\lambda)$ be a strong λ-neighborhood, where $\lambda \in]0,1[$. Then*

(i) *Let $\tau \geq \tau_W$. If $N_\theta(\lambda)$ satisfies (Z1), then $N_\theta(\lambda)$ is convex;*
(ii) *Let $\tau \geq \tau_\Pi$. If $N_\theta(\lambda)$ satisfies (Z2), then $N_\theta(\lambda)$ is convex.*

Proof. (i) Let $\lambda \in]0,1[$ be given. Suppose $p, q \in N_\theta(\lambda)$, we prove that for every $a \in [0,1]$,

$$ap + (1-a)q \in N_\theta(\lambda).$$

If $a = 0$ or $a = 1$, we easily check that $\nu_{ap+(1-a)q} \in N_\theta(\lambda)$, i.e., $ap + (1-a)q \in N_\theta(\lambda)$.

If $\in]0,1[$, since $N_\theta(\lambda)$ satisfies the condition (Z1), one has

$$\nu_{ap}(a\lambda) > 1 - a\lambda \quad \text{and} \quad \nu_{(1-a)q}((1-a)\lambda) > 1 - (1-a)\lambda.$$

Therefore

$$\begin{aligned}
\nu_{ap+(1-a)q}(\lambda) &\geq \tau(\nu_{ap}, \nu_{(1-a)q})(\lambda) \\
&\geq \tau_W(\nu_{ap}, \nu_{(1-a)q})(\lambda) = \sup_{s+t=\lambda} W(\nu_{ap}(s), (1-a)q(t)) \\
&\geq W(\nu_{ap}(a\lambda), \nu_{(1-a)q}((1-\lambda)\lambda)) \\
&> W(1 - a\lambda, 1 - (1-a)\lambda) = 1 - \lambda.
\end{aligned}$$

Thus $ap+(1-a)q$ belongs to $N_\theta(\lambda)$ for every $a \in [0,1]$, i.e., $N_\theta(\lambda)$ is convex.

(ii) It suffices to check that given $p, q \in N_\theta(\lambda)$, with $\lambda \in]0,1[$, then for every $a \in [0,1]$

$$ap + (1-a)q \in N_\theta(\lambda),$$

holds.

If $a = 0$ or $a = 1$, it is obvious that $ap + (1-a)q \in N_\theta(\lambda)$. If $\in]0,1[$, since $N_\theta(\lambda)$ satisfies the condition (Z2), one has

$$\nu_{ap}(a\lambda) > 1 - \frac{a}{\beta_0(a,\lambda)}\lambda$$

and

$$\nu_{(1-\lambda)q}((1-\lambda)\lambda) > 1 - \frac{1-a}{\beta_0(1-a,\lambda)\lambda} = 1 - \frac{1-a}{\beta_0(a,\lambda)\lambda}.$$

One can check that for all $\lambda, a \in]0,1[$

$$1 - \frac{a}{\beta_0(a,\lambda)}\lambda > 0 \quad \text{and} \quad 1 - \frac{1-a}{\beta_0(a,\lambda)}\lambda > 0.$$

Therefore, in view of (ii), one has

$$\begin{aligned}
\nu_{ap+(1-a)q}(\lambda) &\geq \tau(\nu_{ap}, \nu_{(1-a)q})(\lambda) \\
&\geq \tau_\Pi(\nu_{ap}, \nu_{(1-a)q})(\lambda) = \sup_{s+t=\lambda} \Pi(\nu_{ap}(s), (1-a)q(t))
\end{aligned}$$

$$\geq \Pi(\nu_{ap}(a\lambda), \nu_{(1-a)q}((1-\lambda)\lambda))$$
$$> \Pi\left(1 - \frac{a}{\beta_0(a,\lambda)}\lambda, 1 - \frac{1-a}{\beta_0(a,\lambda)\lambda}\right)$$
$$= 1 - \frac{\lambda}{\beta_0(a,\lambda)} + \frac{a(1-a)\lambda^2}{\beta_0(a,\lambda)^2}.$$

Now one proves that

$$1 - \frac{\lambda}{\beta_0(a,\lambda)} + \frac{a(1-a)\lambda^2}{\beta_0(a,\lambda)^2} = 1 - \lambda, \tag{7.3.2}$$

i.e.,

$$\beta_0(a,\lambda)^2 - \beta_0(a,\lambda) + a(1-a)\lambda = 0. \tag{7.3.3}$$

Since $\beta_0(a,\lambda)$ is one root of (7.3.3), (7.3.3) holds and hence (7.3.2) holds as well. Therefore, for all $a \in]0,1[$, $\nu_{ap+(1-a)q}(\lambda) > 1-\lambda$, i.e., $ap+(1-a)q \in N_\theta(\lambda)$, and $N_\theta(\lambda)$ is convex. □

Theorem 7.3.2. *Let (V,ν,τ,τ^*) be a TV PN space and $N_\theta(\lambda)$ be a strong λ-neighborhood of θ, where $\lambda \in]0,1[$.*

(i) *Suppose $\tau \geq \tau_W$. If there is a topologically bounded strong λ-neighborhood $N_\theta(\lambda)$ which satisfies $(Z1)$, then (V,ν,τ,τ^*) is normable.*
(ii) *Suppose $\tau \geq \tau_\Pi$. If there is a topologically bounded strong λ-neighborhood $N_\theta(\lambda)$ which satisfies $(Z2)$, then (V,ν,τ,τ^*) is normable.*

Proof. (i) Let $\lambda \in]0,1[$ be given. Suppose that $N_\theta(\lambda)$ satisfies $(Z1)$ and is topologically bounded. By Theorem 7.3.1(i) the neighborhood is convex. In view of Theorem 6.1.1, it follows that the PN space (V,ν,τ,τ^*) is normable.

(ii) From Theorems 7.3.1(ii) and 7.1.1, by a similar proof to (i), it follows that the PN space (V,ν,τ,τ^*) is normable. □

Theorem 7.3.3. *Let (V,ν,τ,τ^*) be a TV PN space. Let $\overline{N_\theta(\lambda)}$ be the closure of a strong λ-neighborhood $N_\theta(\lambda)$, where $\lambda \in]0,1[$.*

(i) *Suppose $\tau \geq \tau_W$. If there is a strong λ-neighborhood $N_\theta(\lambda)$ which satisfies $(Z1)$, and $\overline{N_\theta(\lambda)}$ is $\mathcal{D}$-compact, then (V,ν,τ,τ^*) is normable.*
(ii) *Suppose $\tau \geq \tau_\Pi$. If there is a strong λ-neighborhood $N_\theta(\lambda)$ which satisfies $(Z2)$, and $\overline{N_\theta(\lambda)}$ is $\mathcal{D}$-compact, then (V,ν,τ,τ^*) is normable.*

Proof. (i) Let N_θ satisfy the assumption (i) of Theorem 7.3.2. From Lemma 6.4.2, it follows that $\overline{N_\theta(\lambda)}$ is topologically bounded. Since $N_\theta(\lambda)$ satisfies $(Z1)$, we know by Theorem 7.3.1(i) that $N_\theta(\lambda)$ is convex. In view of Theorem 7.1.1, one has that the PN space (V, ν, τ, τ^*) is normable.

(ii) In view of Theorems 7.3.1(ii) and 7.1.1, by a similar proof to (i), it follows that the PN space (V, ν, τ, τ^*) is normable. □

The following example shows that there exists a PN space, that is not a Šerstnev space, in which every strong λ-neighborhood $N_\theta(\lambda)$ satisfies (Z1), and we can prove that it is normable by Theorem 7.3.2(i). Firstly we give the following:

Lemma 7.3.1. See (Theorem 3.1(a) in Lafuerza-Guillén *et al.* (1997)). *Let $(V, \|\cdot\|)$ be a normed space and let $\alpha > 1$. If the d.f. $G \in \Delta^+$ is continuous and strictly increasing, then $(V, \|\cdot\|, G, \alpha)$ is a Menger PN space under the strict t-norm defined for all $x, y \in [0, +\infty]$ by*

$$T_G(x, y) := G(\{[G^{-1}(x)]^{1/(1-\alpha)} + [G^{-1}(y)]^{1/(1-\alpha)}\}^{1-\alpha}),$$

where G^{-1} is the inverse of the function G.

Example 7.3.3. Let $G : \mathbb{R} \to [0, 1]$ be defined by $G(x) := \frac{x}{x+1}$. As a consequence: $G(\frac{x}{|p|^3}) = \frac{x}{x+|p|^3}$ for every $p \in \mathbb{R}\backslash\{0\}$, and $\nu : \mathbb{R} \to \Delta^+$ defined by $\nu_0 = \varepsilon_0$ and, for every $p \neq 0$,

$$\nu_p(x) := G\left(\frac{x}{|p|^3}\right).$$

Then the following statements hold:

(i) $(\mathbb{R}, \nu, \tau_T, \tau_{T^*})$ is a Menger PN space under the strict t-norm T defined for all $(s, t) \in [0, 1]^2$ by

$$T(s, t) = G\left(\left[\sqrt{\frac{1-s}{s}} + \sqrt{\frac{1-t}{t}}\right]^{-2}\right);$$

(ii) $(\mathbb{R}, \nu, \tau_T, \tau_{T^*})$ is a TV PN space that is not a Šerstnev space;

(iii) For $\lambda \in]0, \frac{1}{2}[$, every strong λ-neighborhood $N_\theta(\lambda)$ is topologically bounded and satisfies the condition (Z1) in the PN space $(\mathbb{R}, \nu, \tau_T, \tau_{T^*})$;

(iv) $(\mathbb{R}, \nu, \tau_T, \tau_{T^*})$ is normable.

Proof.

(i) According to the definition of G and for $x \in [0,1]$ one has $G^{-1}(x) = \frac{x}{1-x}$. So we have that for $a = 3$ and $(x,y) \in [0,1]^2$ is a Menger PN space under the strict t-norm T defined for all $(x,y) \in [0,1]^2$ by

$$T(x,y) = G\left(\left[\sqrt{\frac{1-x}{x}} + \sqrt{\frac{1-y}{y}}\right]\right),$$

i.e., $T(x,y) = T_G(x,y)$. From Definition 2.4.1 and Lemma 2.4.1 one has that $(\mathbb{R}, \nu, \tau_T, \tau_{T^*})$ is a Menger PN space.

(ii) If $(\lambda_n) \subset \mathbb{R}^+$ satisfies $\lim_{n\to+\infty}(\lambda_n) = 0$, then for all $x \in \mathbb{R}^+$ and $p \neq 0$

$$\lim_{n\to+\infty} \nu_{\lambda_n p}(x) = \lim_{n\to+\infty} G\left(\frac{x}{|\lambda_n p|^3}\right) = \lim_{n\to+\infty} \frac{x}{x + |\lambda_n p|^3} = 1,$$

i.e., $\lim_{n\to+\infty} \nu_{\lambda_n p} = \varepsilon_0$. By Theorem 6.2.8, one has that $(\mathbb{R}, \nu, \tau_T, \tau_{T^*})$ is a TV PN space. Let us see how it is not a Šerstnev space: For $a \notin \{0,1\}$,

$$\nu_{ap}(x) = G\left(\frac{x}{|ap|^3}\right) = \frac{x}{x + |ap|^3} = \frac{x}{x + |a|^3|p|^3},$$

and

$$\nu_p\left(\frac{x}{|a|}\right) = G\left(\frac{\frac{x}{|a|}}{|p|^3}\right) = \frac{\frac{x}{|a|}}{\frac{x}{|a|} + |p|^3} = \frac{x}{x + |a||p|^3}.$$

Thus one has $\nu_{ap}(x) \neq \nu_p(\frac{x}{|a|})$, i.e., the condition (Š) fails.

(iii) Let $\lambda \in]0, \frac{1}{2}[$ be given and suppose

$$\overline{N_\theta(\lambda)} = \{q \in \mathbb{R}; \nu_q(\lambda) \geq 1 - \lambda\}.$$

Suppose any sequence $(p_n) \subset \overline{N_\theta(\lambda)}$, then

$$\nu_{p_n}(\lambda) = G\left(\frac{\lambda}{|p_n|^3}\right) = \frac{\lambda}{\lambda + |p_n|^3} \geq 1 - \lambda.$$

So one has $|p_n| \leq (\frac{\lambda^2}{1-\lambda})^{\frac{1}{3}}$. Thus (p_n) has a classical convergent subsequence (p_{n_k}) in $\mathbb{R}$. Therefore (p_{n_k}) is convergent to $p \in \mathbb{R}$, i.e.,

$(p_{n_k} - p) \to 0$ with $n_k \to \infty$, and p satisfies $|p| \le (\frac{\lambda^2}{1-\lambda})^{\frac{1}{3}}$. Hence for any $x > 0$,

$$\lim_{n_k\to\infty} \nu_{(p_{n_k}-p)}(x) = \lim_{n_k\to\infty} G\left(\frac{x}{|p_{n_k}-p|^3}\right) = \lim_{n_k\to\infty} \frac{x}{x+|p_{n_k}-p|^3} = 1,$$

and

$$\nu_p(\lambda) = G\left(\frac{\lambda}{|p|^3}\right) = \frac{\lambda}{\lambda+|p|^3} \ge 1-\lambda,$$

i.e., the sequence (p_n) in $\overline{N_\theta(\lambda)}$ has a convergent subsequence $(p_{n_k}$ convergent to an element $p \in \overline{N_\theta(\lambda)}$. From Definition 3.1.1(iii), we know that $\overline{N_\theta(\lambda)}$ is $\mathcal{D}$-compact. And by Lemma 6.4.1, $\overline{N_\theta(\lambda)}$ is topologically bounded.

Now one shows that in the above PN space $(\mathbb{R}, \nu, \tau_T, \tau_{T^*})$, every strong λ-neighborhood $N_\theta(\lambda)$ satisfies condition (Z1) for $\lambda \in]0, \frac{1}{2}[$. In fact, let $\lambda \in]0, \frac{1}{2}[$ be given, if $N_\theta(\lambda)$ does not satisfy the condition (Z1), i.e., $\nu_p(\lambda) > 1-\lambda$, but $\nu_{ap}(a\lambda) \le 1 - a\lambda$ for $a \in]0,1[$, then

$$\nu_p(\lambda) = G\left(\frac{\lambda}{|p|^3}\right) = \frac{\lambda}{\lambda+|p|^3} \ge 1-\lambda \Rightarrow |p|^3 < \frac{\lambda^2}{1-\lambda},$$

and

$$\nu_{ap}(a\lambda) = G\left(\frac{a\lambda}{|ap|^3}\right) = \frac{a\lambda}{a\lambda+|ap|^3} \le 1-a\lambda \Rightarrow |p|^3 \ge \frac{\lambda^2}{a(1-a\lambda)}.$$

Now

$$\frac{\lambda^2}{a(1-a\lambda)} \ge \frac{\lambda^2}{1-\lambda} \Leftrightarrow \lambda \le \frac{1}{1+a}.$$

From $a \in]0,1[$, one has that $\frac{1}{1+a} > \frac{1}{2}$. Thus for $\lambda \in]0, \frac{1}{2}[$, one has

$$|p|^3 \ge \frac{\lambda^2}{a(1-a\lambda)} \ge \frac{\lambda^2}{1-\lambda} > |p|^3.$$

This leads to a contradiction, and hence every strong λ-neighborhood $N_\theta(\lambda)$ satisfies condition (Z1) for $\lambda \in]0, \frac{1}{2}[$.

(iv) In view of (ii), (iii), and Theorem 6.3.2(i), the PN space $(\mathbb{R}, \nu, \tau_T, \tau_{T^*})$ is normable. □

Example 7.3.4. Let $(V, |\cdot|)$ be a normed space. If $\nu : V \to \Delta^+$ is given by

$$\nu_p(x) = \begin{cases} 0, & x \le 0 \\ exp(-\|p\|^2), & 0 < x < +\infty \\ 1, & x = +\infty). \end{cases}$$

for every $p \in V$. Let $N_\theta(\lambda)$ be a strong λ-neighborhood in the PN space $(V, \nu, \tau_\Pi, \tau_{\Pi^*})$. Then

(i) $(V, \nu, \tau_\Pi, \tau_{\Pi^*})$ is a Menger PN space satisfying the properties $(F1)$ to $(F4)$ of Theorem 3.6.
(ii) Let $\lambda_1 = 1 - e^{-\frac{1}{2}}$, then for all $\lambda \in]0, \lambda_1]$, $N_\theta(\lambda)$ satisfies $(Z2)$.

Proof.

(i) Suppose $T = \Pi$ and $f(x) = exp(-x^2)$ in Theorem 6.2.2, one has that the hypothesis of Example 6.3.4 satisfy all the hypothesis of Theorem 6.2.2. Therefore $(V, \nu, \tau_\Pi, \tau_{\Pi^*})$ is a Menger PN space satisfying the properties $(F1)$ to $(F4)$.
(ii) From

$$\nu_p(\lambda) = exp(-\|p\|^2) > 1 - \lambda \Leftrightarrow \|p\|^2 < -\ln(1-\lambda)$$

and

$$\begin{aligned} \nu_{ap}(a\lambda) &= exp(-\|ap\|^2) > 1 - \frac{a}{\beta_0(a,\lambda)}\lambda \\ &\Leftrightarrow \|ap\|^2 < -\ln\left(1 - \frac{2a\lambda}{1+\sqrt{1-4a(1-a)\lambda}}\right), \end{aligned}$$

Suppose that for $a \in]0, 1[$,

$$-a^2 \ln(1-\lambda) < -\ln\left(1 - \frac{2a\lambda}{1+\sqrt{1-4a(1-a)\lambda}}\right), \tag{7.3.4}$$

then $N_\theta(\lambda)$ satisfies (Z2).

One checks that (6.3.4) holds if, and only if, the following (6.3.5) holds

$$\begin{aligned} &\text{For all } a \in]0,1[, \text{one has } (1-\lambda)^{a^2} \\ &> 1 - \ln\left(1 - \frac{2a\lambda}{1+\sqrt{1-4a(1-a)\lambda}}\right). \end{aligned} \tag{7.3.5}$$

Now one proves that (6.3.5) holds. Firstly one considers that the following (6.3.6) holds:

$$\text{For all} \quad a \in]0,1[\qquad (1-\lambda)^{a^2} > 1 - a\lambda. \tag{7.3.6}$$

In fact, let $g(a) = (1-\lambda)^{a^2} - (1-a\lambda)$. From $g'(a) = 2a\ln(1-\lambda)(1-\lambda)^{a^2} + \lambda$, one has

$$\begin{aligned} g''(a) &= 4a^2[\ln(1-\lambda)]^2(1-\lambda)^{a^2} + 2\ln(1-\lambda)(1-\lambda)^{a^2} \\ &= 2\ln(1-\lambda)(1-\lambda)^{a^2}(1+2a^2\ln(1-\lambda)). \end{aligned}$$

Since for $\lambda \in]0, \lambda_1[$,

$$\begin{aligned} 1 + 2a^2 \ln(1-\lambda) &\geq 1 + 2a^2 \ln(1-\lambda_1) = 1 + \ln(1-\lambda_1)^{2a^2} \\ &= 1 + \ln(e^{-a^2}) = 1 - a^2 > 0. \end{aligned}$$

Thus $g''(a) < 0$, i.e. $g(a)$ is a convex function for $a \in]0,1[$. It is easy to check that $g(0) = 0$ and $g(1) = 0$. Therefore for $a \in]0,1[$ one has that $g(a) > 0$, i.e., (7.3.6) holds. Because for every $a \in]0,1[$

$$\begin{aligned} &(1 - a\lambda) - \left(1 - \frac{2a\lambda}{1 + \sqrt{1 - 4a(1-a)\lambda}}\right) \\ &= \frac{2a\lambda}{1 + \sqrt{1 - 4a(1-a)\lambda}} - a\lambda \\ &= a\lambda \left(\frac{2}{1 + \sqrt{1 - 4a(1-a)\lambda}} - 1\right) \\ &= a\lambda \left(\frac{1 - \sqrt{1 - 4a(1-a)\lambda}}{1 + \sqrt{1 - 4a(1-a)\lambda}}\right) \\ &> 0, \end{aligned}$$

i.e.,

$$\text{for every} \quad a \in]0,1[, \quad (1 - a\lambda) > \left(1 - \frac{2a\lambda}{1 + \sqrt{1 - 4a(1-a)\lambda}}\right). \tag{7.3.7}$$

By (7.3.6) and (7.3.7), one has that (7.3.5) holds. Therefore for $\lambda \in]0, \lambda_1]$, $N_\theta(\lambda)$ satisfies (Z2). □

Remark 7.3.2. In view of Example 7.3.4, ($F3$) and Theorem 7.3.4(ii), one has that the PN space $(V, \nu, \tau_\Pi, \tau_{\Pi^*})$ is normable, which is consistent with ($F2$) of Theorem 11 in Lafuerza-Guillén *et al.* (2008).

Theorem 7.3.4. *Let (V, ν, τ, τ^*) be a finite dimensional PN space, where τ^* is Archimedean, $\nu_p \neq \varepsilon_\infty$, $\nu(V) \subseteq \mathcal{D}^+$ and $\mathcal{D}^+$ is invariant under τ, for every $p \in V$ on the real field $(\mathbb{R}, \nu', \tau', \tau'^*)$, where ν' has the LG-property. Then the following statements hold.*

(i) *Suppose $\tau \geq \tau_W$. If there is a strong λ-neighborhood $N_\theta(\lambda)$ which is $\mathcal{D}$-bounded and satisfies ($Z1$) for some $\lambda \in]0,1]$, then (V, ν, τ, τ^*) is normable.*

(ii) *Suppose $\tau \geq \tau_{\Pi}$. If there is a strong λ-neighborhood $N_\theta(\lambda)$ which is $\mathcal{D}$-bounded and satisfies (Z2) for some $\lambda \in]0,1]$, then (V, ν, τ, τ^*) is normable.*

Proof. Since τ^* is Archimedean, by Theorem 3.3.3 the PN space (V, ν, τ, τ^*) is a TV PN space.

(i) Let $N_\theta(\lambda)$ satisfy the hypothesis of Theorem 7.3.4(i). By Theorem 7.3.2(i) one has that $N_\theta(\lambda)$ is topologically bounded. From Theorem 7.3.2(i) the PN space (V, ν, τ, τ^*) is normable.

(ii) In view of Theorem 7.3.4 and 7.3.2(ii), by the similar proof of (i), the statement (ii) holds. □

7.4 Open Questions

In the previous sections we have been able to characterize:

(a) Those Šerstnev PN spaces that are normable topological vector spaces;

(b) A large class of Menger PN spaces, i.e., those α-simple PN spaces under the strict t-norm T defined for all $x, y \in [0, +\infty]$ by

$$T_G(x, y) := G(\{[G^{-1}(x)]^{1/(1-\alpha)} + [G^{-1}(y)]^{1/(1-\alpha)}\}^{1-\alpha}),$$

where G^{-1} is the inverse of the function G , which are not Šerstnev, and that are normable topological vector spaces.

Several questions remain open:

(i) To give at least sufficient conditions under which a general PN space of the type $(V, \nu, \tau_{T_1}, \tau_{T_2})$, not necessarily a Menger PN space, with $T_1 \leq T_2$, is normable.

(ii) To give at least sufficient conditions under which a general PN space of the type $(V, \nu, \tau_{T_1,L_2}, \tau_{T_2,L_1})$, with $T_1 \leq T_2$ and $L_1 \leq L_2$, is normable, i.e., general Menger PN spaces.

(iii) To give at least sufficient conditions under which a general PN space is normable; more, to characterize (rather than just having a sufficient condition) the class of PN spaces that are also topological vector spaces, and, once this has been achieved, to study normability in the class thus determined.

Chapter 8

Invariant and Semi-Invariant PN Spaces

The aim of this chapter is to introduce concepts such as invariance and semi-invariance among the PN spaces (see Gahemi, Lafuerza-Guillén, and Saiedinezhad, 2009) and identify a sufficient condition for some PN spaces to be semi-invariant. Urysohn's lemma and the Tietze extension theorem for PN spaces are established in this chapter.

8.1 Invariance and Semi-Invariance

In any PN space (V, ν, τ, τ^*) with regard to the distance d_S, one has, in general, the following:

$$d_S(\nu_{p-q}, \varepsilon_0) \neq d_S(\nu_p, \nu_q).$$

But it would be interesting to know in which cases and under which conditions the inequality $d_S(\nu_{p-q}, \varepsilon_0) \neq d_S(\nu_p, \nu_q)$ holds. In this case, the corresponding relationship in functional analysis is

$$\|p - q\| \geq \|p\| - \|q\|.$$

Definition 8.1.1. The probabilistic normed space (V, ν, τ, τ^*) is said to be *invariant*, if for every $p, q \in V$ we have $d_S(\nu_{p-q}, \varepsilon_0) = d_S(\nu_p, \nu_q)$.

Definition 8.1.2. The probabilistic normed space (V, ν, τ, τ^*) is said to be *semi-invariant*, if for every $p, q \in V$ we have $d_S(\nu_{p-q}, \varepsilon_0) > d_S(\nu_p, \nu_q)$.

Example 8.1.1. The quadruple $(\mathbb{R}, \nu, \tau_\pi, \tau_{\pi^*})$ where $\nu : \mathbb{R} \to \Delta^+$ is defined by

$$\nu_p(x) := \begin{cases} 0, & \text{if } x = 0 \\ exp\,(-\sqrt{\|p\|}), & \text{if } 0 < x < +\infty, \\ 1, & \text{if } x = +\infty \end{cases}$$

and $\nu_0 = \varepsilon_0$ is strictly a semi-invariant PN space.

Proof. $(\mathbb{R}, \nu, \tau_\pi, \tau_{\pi^*})$ is a Menger PN space that is not a Šerstnev space. We have

$$\begin{aligned} d_S(\nu_{p-q}, \varepsilon_0) &= \inf\{h \in]0,1[: \nu_{p-q}(h^+) > 1-h\} \\ &= \inf\{h \in]0,1[: exp\,(\|p-q\|^{1/2}) > 1-h\} \\ &= 1 - exp\,(\|p-q\|^{1/2}). \end{aligned}$$

Suppose without loss of generality $\|p\| > \|q\|$. Then for every $h \in]0,1[$ and for all $t > 0$, particularly for $t \in]0, 1/h[$, one has

$$\nu_p(t) \leq \nu_q(t+h) + h$$

so that, in our example, the condition $[\nu_p, \nu_q; h]$ says that

$$h \geq exp\,(-\sqrt{\|q\|}) - exp\,(-\sqrt{\|p\|}).$$

Consequently, if $p \neq q$,

$$d_S(\nu_p, \nu_q) = exp\,(-\sqrt{\min\{\|q\|, \|p\|\}}) - exp\,(-\sqrt{\max\{\|q\|, \|p\|\}})$$

holds. And taking into account the following relations

$$\begin{aligned} \sqrt{\|p\|} &\leq \sqrt{\|p-q\| + \|q\|} \leq \sqrt{\|p-q\|} + \sqrt{\|q\|} \\ &\Rightarrow exp\,(-\sqrt{\|p\|}) \leq exp\,(-\sqrt{\|p-q\|}).exp\,(-\|\sqrt{q}\|) \\ &\Rightarrow exp\,(-\|\sqrt{q}\|) - exp\,(-\|\sqrt{p}\|) \leq exp\,(-\|\sqrt{q}\|) \\ &\quad -exp\,(-\sqrt{\|p-q\|}).exp\,(-\|\sqrt{q}\|) \\ &= exp\,(-\|\sqrt{q}\|).[1 - exp\,(-\sqrt{\|p-q\|})] \leq 1 - exp\,(-\sqrt{\|p-q\|}) \end{aligned}$$

one has finally the strict inequality $d_S(\nu_{p-q}, \varepsilon_0) > d_S(\nu_p, \nu_q)$. □

Lemma 8.1.1. *If $a, b \in [0, +\infty]$ then the statement*

$$d_S(\varepsilon_a, \varepsilon_b) = \min\left\{1, \frac{1}{\min\{a,b\}}, |a-b|\right\}$$

holds. Particularly, the cases $b = 0$ and $b = +\infty$ are

$$d_S(\varepsilon_a, \varepsilon_0) = \min\{1, a\}$$

$$d_S(\varepsilon_a, \varepsilon_\infty) = \min\left\{1, \frac{1}{a}\right\}$$

Proof. Based on the definition of the Sibley metric, one has

$$d_S(\varepsilon_a, \varepsilon_b) = \inf\{h \in]0,1[: \text{ both } [\varepsilon_a, \varepsilon_b, h] \text{ and } [\varepsilon_b, \varepsilon_a, h], \text{ hold } \}.$$

Let us recall that for every $x \in]0, \frac{1}{h}[$ the relation $[\varepsilon_a, \varepsilon_b, h] \Leftrightarrow \varepsilon_b(x) \leq \varepsilon_a(x+h)+h$ holds.

Consequently, for all $x \in]0, \frac{1}{h}[$

$$d_S(\varepsilon_a, \varepsilon_b) = \inf\{h \in]0,1[: \varepsilon_{\min\{a,b\}}(x) \leq \varepsilon_{\max\{a,b\}}(x+h)+h\} \quad (8.1.1)$$

The inequality on the right-hand side of (8.1.1) occurs in the following cases:

(a) $h > 1$;
(b) $\frac{1}{h} \leq \min\{a,b\}$ or equivalently $h \geq \frac{1}{\min\{a,b\}}$;
(c) For every $x \in]\min\{a,b\}, 1/h[$ one has

$$1 \leq \varepsilon_b(x) \leq \varepsilon_a(x+h)+h \quad \text{and} \quad h < \frac{1}{\min\{a,b\}}$$

$$\Leftrightarrow h < \min\left\{1, \frac{1}{\min\{a,b\}}\right\}$$

$$\text{with} \quad x+h > \max\{a,b\} \quad \text{for all } x \in]\min\{a,b\}, 1/h[$$

$$\Leftrightarrow h < \min\left\{1, \frac{1}{\min\{a,b\}}\right\}$$

$$\text{with } h > \max\{a,b\} - \min\{a,b\} = |a-b|$$

for all $x \in]\min\{a,b\}, 1/h[$. □

Theorem 8.1.1. *Let (V, ν, τ, τ^*) be a PN space where V is a linear space $\nu_p = \varepsilon_{\varphi(p)}$ and $\varphi : V \to \mathbb{R}$ is a positive function such that for every $p, q \in V : \varphi(q) - \varphi(p) \leq \varphi(q-p)$. Then (V, ν, τ, τ^*) is a semi-invariant PN space.*

Proof. It is enough, if we prove the following relation

$$d_S(\varepsilon_{\varphi(q)}, \varepsilon_{\varphi(p)}) \leq d_S(\varepsilon_{\varphi(q-p)}, \varepsilon_0).$$

Let p, q be in V. Then by Lemma (8.1.1), one has:

$$\begin{aligned} d_S(\nu_{p-q}, \varepsilon_0) &= d_S(\varepsilon_{\varphi(q-p)}, \varepsilon_0) = \min\{1, \varphi(p-q)\} \\ &\geq \min\left\{1, |\varphi(p) - \varphi(q)|, \frac{1}{\min\{\varphi(p), \varphi(q)\}}\right\} \\ &= d_S(\varepsilon_{\varphi(q)}, \varepsilon_{\varphi(p)}) = d_S(\nu_p, \nu_q). \end{aligned}$$

□

Example 8.1.2. The quadraple (V, ν, τ_M, Π_M) where $(V, \|\cdot\|)$ is a classical normed space and $\nu_p = \varepsilon_{\varphi(p)}$ with $\varphi(p) = \frac{\|p\|}{1+\|p\|}$, is a semi-invariant PN space.

Proof. (V, ν, τ_M, Π_M) is a PN space, which is not a Šerstnev space. From Lemma (8.1.1) $d_S(\varepsilon_{\varphi(p-q)}, \varepsilon_0) = \frac{\|p-q\|}{1+\|p-q\|}$. On the other hand the function $\varphi(p) = \frac{\|p\|}{1+\|p\|}$ satisfies the condition given in Theorem (8.1.1)

$$\begin{aligned}\varphi(q) - \varphi(p) &= \frac{\|q\|}{1+\|q\|} - \frac{\|p\|}{1+\|p\|} \\ &= \frac{\|q\| - \|p\|}{1+\|q\|+\|p\|+\|p\|\|q\|} \\ &\leq \frac{\|q\| - \|p\|}{1+\|q\|+\|p\|} \leq \frac{\|q-p\|}{1+\|q-p\|} = \varphi(q-p)\end{aligned}$$

and the proof is complete. □

It is possible to give sufficient conditions under which the inequality

$$d_S(\nu_{p-q}, \varepsilon_0) \geq d_S(\nu_p, \nu_q)$$

holds, as in the following theorem, but first we need a lemma:

Lemma 8.1.2. *Let $F_i, G_i (i = 1, 2)$ be the distance distribution functions in Δ^+. Let $A_i; i = 1, 2$ be the set defined via*

$$A_i := \inf\{h \in]0, 1[: \text{ both } [F_i, G_i, h] \text{ and } [G_i, F_i, h], \text{ hold}\}$$

Then

$$\begin{aligned}d_S(F_1, G_1) < d_S(F_2, G_2) &\quad \text{if and only if } A_1 \supset A_2 \\ d_S(F_1, G_1) = d_S(F_2, G_2) &\quad \text{if and only if } A_1 = A_2.\end{aligned}$$

Proof. Let us recall that $d_S(F_i, G_i) = \inf A_i (i = 1, 2)$. By the definition of A_i, if $h_0 \in A_i$ then $]h_0, 1[\subset A_i \subset]0, 1[$; Therefore, if $h = \inf A_i > 0$, then

$$[F_i, G_i, h] \quad \text{and} \quad [G_i, F_i, h]$$

hold.

Moreover

$$A_i := \begin{cases}]0, 1[, & \text{if } d_S(F_i, G_i) = 0 \\ [a_i, 1], & \text{if } d_S(F_i, G) = a_i. \end{cases}$$

□

Theorem 8.1.2. *Let C be an associative copula and (V, ν, τ_C, τ^*) a PN space; then (V, ν, τ_C, τ^*) is semi-invariant.*

Proof. Since C is an associative copula, then C is a continuous t-norm and as a consequence one may say that τ_C is a continuous triangle function.

When calculating $d_S(\nu_{p-q}, \varepsilon_0)$ note that the condition $[\varepsilon_0, \nu_{p-q} : h]$ is always right. The other one, $[\nu_{p-q}, \varepsilon_0 : h]$, tells us that given $h \in]0,1[$ and for all $x \in]0, 1/h[$, the relation $1 \leq \nu_{p-q}(x+h) + h$ holds. From Lemma (8.1.2) assume that $F_1 = \nu_p, G_1 = \nu_q$ and $F_2 = \nu_{p-q}, G_2 = \varepsilon_0$. We have to prove that given $h \in]0,1[$ and for all $x \in]0, 1/h[$, if

$$1 - h \leq \nu_{p-q}(x+h), \tag{8.1.2}$$

then

$$\nu_p(x) \leq \nu_q(x+h) + h \quad \text{and} \quad \nu_q(x) \leq \nu_p(x+h) + h.$$

hold. We only prove the second inequality because the other one is symmetrical: it suffices to interchange p, q.

For every $x \in]0, 1/h[$ and applying (8.1.2) one has

$$\begin{aligned}
\nu_p(x+h) + h &\geq \tau_C(\nu_{p-q}, \nu_q)(x+h) + h \\
&= \sup\{C(\nu_{p-q}(u), \nu_q(v)) : u + v = x + h\} + h \\
&\geq \sup\{C(\nu_{p-q}(u), \nu_q(v)) : u + v = x + h, \nu\langle x, u\rangle h\} + h \\
&\geq \sup\{C(1-h, \nu_q(v)) : \nu < x\} + h \\
&= C(1-h, \nu_q(x)) + h.
\end{aligned}$$

Moreover, since C is a copula, it follows that

$$\begin{aligned}
C(1,1) - C(1, \nu_q(x)) - C(1-h, 1) + C(1-h, \nu_q(x)) \\
= 1 - \nu_q(x) - (1-h) + C(1-h, \nu_q(x)) \geq 0,
\end{aligned}$$

so that

$$C(1-h, \nu_q(x)) + h \geq \nu_q(x).$$

□

As a consequence of this theorem, if (V, ν, τ_T, τ^*) is a PN space such that $\tau_T \geq \tau_C$ with C an associative copula, then (V, ν, τ_T, τ^*) is semi-invariant.

Corollary 8.1.1.

(a) *Every equilateral space (V, F, Π_M) is semi-invariant;*
(b) *Every simple space $(V, \|\cdot\|, G, M)$ is semi-invariant;*
(c) *Every PN space (S, ν) is semi-invariant.*

8.2 New Class of PN Spaces

Before introducing a new class of PN spaces we need the following technical Lemma (see Lafuerza-Guillén, Rodríguez Lallena, and Sempi (1999)).

Lemma 8.2.1. *Let $f : [0, +\infty] \to [0, 1]$ be a right-continuous, non-increasing function. Let us define $f^{-1}(1) = 0$ and $f^{-1}(y) := \sup\{x; f(x) > y$ for all $y \in [0, 1[\}$ ($f^{-1}(y)$ might be infinite). If $x_0 \in [0, +\infty]$ and $y_0 \in [0, 1]$, then the followings facts are equivalent:*

(a) $f(x_0) > y_0$;
(b) $x_0 < f^{-1}(y_0)$.

Proof. If $f(x_0) > y_0$ then $f^{-1}(y_0) = \sup\{x; f(x) > y_0\} \geq x_0$. If we suppose that $\sup\{x; f(x) > y_0\} = x_0$ then $f(x_0) \leq y_0$ for every $x > x_0$. Thus $f(x_0) = f(x_0+) \leq y_0$, against the assumption; whence $(a) \Rightarrow (b)$. The converse result is an immediate consequence of the monotonicity of f. □

The following theorem introduces a new class of PN spaces which generalizes an example (see Lafuerza-Guillén (1996)) and also provides some properties of the spaces in that class.

Theorem 8.2.1. *Let $(V, \|\cdot\|)$ be a normed space and let T be a continuous t-norm. Let f be a function as in Lemma (8.2.1), and satisfying the following two properties:*

(a) *$f(x) = 1$ if and only if $x = 0$;*
(b) *$f(\|p + q\|) \geq T(f(\|p\|), f(\|q\|))$ for every $p, q \in V$.*

If $\nu : V \to \Delta^+$ is given by

$$\nu_p(x) = \begin{cases} 0, & \text{if } x < 0 \\ f(\|p\|), & \text{if } x \in]0, +\infty[\\ 1, & \text{if } x = +\infty \end{cases}$$

for every $p \in V$, then (V, ν, τ_T, τ_M) is a Menger PN space satisfying the following properties:

(F1) *(V, ν, τ_T, τ_M) is a TV space;*
(F2) *(V, ν, τ_T, τ_M) is normable;*
(F3) *If $p \in V$ and $t > 0$, then the strong neighborhood $N_p(t)$ in (V, ν, τ_T, τ_M) is not $\mathcal{D}$-bounded, but $N_p(t)$ is topologically bounded whenever $N_p(t) \neq V$;*
(F4) *(V, ν, τ_T, τ_M) is not a Šerstnev space;*
(F5) *(V, ν, τ_T, τ_M) is not a characteristic PN space.*

Now we consider some special cases and use the preceding theorem in order to give some examples.

Example 8.2.1. Suppose that, in Theorem (8.2.1), $T = \Pi$. Then, property (b) reads $f(\|p+q\|) \geq f(\|p\|)f(\|q\|)$ for all $p, q \in V$. It is not difficult to prove that, under the given assumptions on f, property (b) is equivalent to the following:

$$f(x+y) \geq f(x)f(y), \quad \text{for all} \quad x, y \in]0, \infty[. \tag{8.2.1}$$

The following are examples of functions f satisfying the assumptions of Theorem (8.2.1):

$$f_{\alpha,\beta}(x) := 1 - \frac{\beta}{\alpha} + \frac{\beta}{x+\alpha}, \quad 0 \leq \beta \leq \alpha;$$
$$g_{\alpha,\beta}(x) := 1 - \alpha + \alpha exp\,(-x^{\beta}), \quad 0 < \beta \leq 1, \ \beta > 0.$$

Example 8.2.2. Take $T = W$ in Theorem (8.2.1). In this case property (b) reads

$$f(\|p+q\|) \geq f(\|p\|)f(\|q\|) - 1 \ \text{ for all } \ p, q \in V.$$

Since W is the smallest continuous t-norm, all the functions f satisfying the assumptions of Theorem 8.2.1 with respect to any t-norm T also satisfy such assumptions with respect to W. It is not hard to prove that, under those assumptions, property (b) is equivalent to the following:

$$1 + f(x+y) \geq f(x) + f(y) \ \text{ for all } \ x, y \in]0, \infty[.$$

For instance, the following functions satisfy this property but not that considered in Example (8.2.1), since they do not satisfy (8.2.1):

$$h_{\alpha,\beta}(x) := \begin{cases} 1 - \alpha x, & \text{if } 0 \leq x \leq \beta \\ 1 - \alpha\beta, & \text{if } x > \beta \end{cases}$$
$$0 < \beta \leq 1/\alpha$$

Theorem 8.2.2. *Every PN space belonging to the class considered in Theorem* (8.2.1) *is semi-invariant.*

Proof. Since $d_S(\nu_{p-q}, \varepsilon_0) = \inf\{h \in]0,1[: \nu_{p-q}(h^+) > 1 - h\}$ one has $f(\|p-q\|) > 1 - h \Rightarrow h > 1 - f(\|p-q\|)$, it follows that:

$$d_S(\nu_{p-q}, \varepsilon_0) = 1 - f(\|p-q\|).$$

On the other hand,

$$d_S(\nu_p, \nu_q) = \inf\{h \in]0,1[: \text{ both } [\nu_p, \nu_q, h] \text{ and } [\nu_q, \nu_p, h], \text{ hold}\}.$$

Suppose, without loss of generality, $\|p\| \geq \|q\|$, then $[\nu_q, \nu_p, h]$ is equivalent to $f(\|p\|) \leq f(\|q\|) + h$. In fact, this inequality is strict. Moreover, from

$[\nu_p, \nu_q, h]$ one has $h \geq f(\|q\|) - f(\|p\|)$, hence

$$d_S(\nu_p, \nu_q) = f(\|q\|) - f(\|p\|).$$

Now we need to investigate under which particular conditions one has, for the PN spaces considered in Theorem (8.2.1), the inequality

$$1 - f(\|p - q\|) \geq f(\|q\|) - f(\|p\|). \tag{8.2.2}$$

If one chooses f among the type's functions $f_{\alpha,\beta}$, then it is not difficult to check that

$$\frac{\|p\| - \|q\|}{\|p\|\|q\| + (\|p\| + \|q\|)\alpha + \alpha^2} \leq \frac{\|p - q\|}{\alpha^2 + \|p - q\|\alpha}.$$

and that the inequality $d_S(\nu_{p-q}, \varepsilon_0) \geq d_S(\nu_p, \nu_q)$ holds. And if one chooses f among the type's functions $g_{\alpha,\beta}$, then the PN spaces of Theorem (8.2.1) are always strictly semi-invariant: since

$$d_S(\nu_{p-q}, \varepsilon_0) = 1 - g_{\alpha,\beta}(\|p - q\|) = \alpha - \alpha^{\|p-q\|^\beta},$$

and

$$d_S(\nu_p, \nu_q) = g_{\alpha,\beta}(\|q\|) - g_{\alpha,\beta}(\|p\|) = \alpha e^{-\|q\|^\beta} - \alpha e^{-\|p\|^\beta},$$

one has only to check the inequality

$$1 - e^{-\|p-q\|^\beta - 1} \geq e^{-\|q\|^\beta}(1 - e^{-(\|p\|^\beta - \|q\|^\beta)}),$$

which is equivalent to the inequality

$$1 - e^{-\|p-q\|^\beta} \geq 1 - e^{-(\|p\|^\beta - \|q\|^\beta)},$$

and this is true because of the well-known inequality $(1 - s)^\beta \geq 1 - s^\beta$. In fact, the inequality is strict and the verification is complete. Finally, if one chooses f among the type's functions $h_{\alpha,\beta}$ the PN spaces considered in Theorem 8.2.1 are also semi-invariant. □

Definition 8.2.1. (Wilansky, 1964) A topological space is called *normal space*, if any two disjoint closed subset of it can be seperated by open sets.

Theorem 8.2.3. *Every semi-invariant PN space (V, ν, τ, τ^*) is normal.*

Proof. If A, B are a closed subset of V we should construct two disjoint subsets U, W of V such that $A \subset U$ and $B \subset W$.

Let $U := \{p \in V; \inf_{a \in A} d_S(\nu_{p-a}, \varepsilon_0) < \inf_{b \in B} d_S(\nu_{p-b}, \varepsilon_0)\}$. If $p \in A$ then $\inf_{a \in A} d_S(\nu_{p-a}, \varepsilon_0) = 0$ and hence $A \subset U$. Now we prove that U is open.

Suppose $p \in U, \alpha_p = \inf d_S(\nu_{p-a}, \varepsilon_0)$ and $\beta_p = \inf d_S(\nu_{p-b}, \varepsilon_0)$. Therefore, $\alpha_p < \beta_p$.

For every $\varepsilon > 0$ there exists $a_0 \in A$ such that $d_S(\nu_{p-a_0}, \varepsilon_0) < \alpha_p + \varepsilon$.

For every $p \in U$ we will show that $N_p(\frac{\beta_p - \alpha_p}{2}) \subset U$.

Let $q \in N_p(\frac{\beta_p - \alpha_p}{2})$ then $d_S(\nu_{p-q}, \varepsilon_0) < \frac{\beta_p - \alpha_p}{2}$ and

$$\begin{aligned}\inf d_S(\nu_{q-a}, \varepsilon_0) &\le \inf d_S(\nu_{q-a}, \nu_{q-p}) + d_S(\nu_{q-p}, \varepsilon_0)\\ &\le \inf d_S(\nu_{q-a}, \nu_{q-p}) + \frac{\beta_p - \alpha_p}{2}.\end{aligned}$$

The space (V, ν, τ, τ^*) is semi-invariant and hence

$$\inf d_S(\nu_{q-a}, \nu_{q-p}) \le \inf d_S(\nu_{p-a}, \varepsilon_0) = \alpha_p.$$

Therefore, $d_S(\nu_{q-a}, \varepsilon_0) \le \alpha_p + \frac{\beta_p - \alpha_p}{2} = \frac{\beta_p + \alpha_p}{2} < \beta_p$ and $q \in U$.

Similarly, if $W := \{p \in V; \inf d_S(\nu_{p-a}, \varepsilon_0) > \inf d_S(\nu_{p-b}, \varepsilon_0)\}$, then W is an open subset of V such that $B \subset W$, and the construction of U and W shows that $U \cap W = \phi$. □

Immediately the next corollaries come from Theorem (8.2.3).

Corollary 8.2.1. (*Urysohn's lemma for PN space*) *Any two disjoint subsets of every semi-invariant PN space can be separated by a continuous function.*

Corollary 8.2.2. (*The Tietze extension theorem for PN space*) *If A is any closed subset of the semi-invariant PN space (V, ν, τ, τ^*), and $f \in C(A, [a, b])$, then there exists $F \in C(V, [a, b])$ such that $F|_A = f$.*

Corollary 8.2.3. *If A is any closed subset of the semi-invariant PN space (V, ν, τ, τ^*) and $f \in C(A)$ then there exists $F \in C(V)$ such that $F|_A = f$.*

8.3 Open Questions

(1) Is any α-simple space invariant or semi-invariant? If not, are there any conditions under which it is? Also, if not, why not?

Chapter 9

Linear Operators

9.1 Boundedness of Linear Operators

Some of the classical results still hold, with proofs that are similar to the usual ones, except for a change of language. For instance, one has the following:

Theorem 9.1.1. *Given two PN spaces (V_1, ν, τ, τ^*) and $(V_2, \mu, \sigma, \sigma^*)$, a linear map $T : V_1 \to V_2$ is either continuous at every point of V_1 or at no point of V_1. In particular, T is continuous if, and only if, it is continuous at θ_1, the null vector of V_1.*

We recall that, in general, an operator T from a metric or normed space V into another metric or normed space V' is said to be bounded if it maps every bounded subset A of V into the bounded subset $T(A)$ of V'. It is therefore interesting to adopt the terminology introduced in the previous section in order to classify linear operators with respect to boundedness properties.

Definition 9.1.1. A linear map T between the PN spaces (V_1, ν, τ, τ^*) to $(V_2, \mu, \sigma, \sigma^*)$ is said to be

(a) *certainly bounded* if it maps every certainly bounded set A of the space (V_1, ν, τ, τ^*) into certainly bounded set $T(A)$ of the space $(V_2, \mu, \sigma, \sigma^*)$, i.e., if there exists $t_0 \in]0, +\infty[$ such that $\nu_p(t_0) = 1$ for every $p \in A$, then there exists $t_1 \in]0, +\infty[$ such that $\mu_{T_p}(t_1) = 1$ for every $p \in A$;

(b) *bounded*, if it maps every $\mathcal{D}$-bounded set of V_1 into a $\mathcal{D}$-bounded set of V_2, i.e., if R_{TA} belongs to $\mathcal{D}^+$ for every $\mathcal{D}$-bounded set A of V_1. Equivalently, T is bounded if the implication

$$\lim_{t \to +\infty} \varphi_A(t) = 1 \Longrightarrow \lim_{t \to +\infty} \varphi_{TA}(t) = 1.$$

is satisfied for every non-empty subset A of V_1;

(c) *strongly **B**-bounded*, if there exists a constant $k > 0$ such that, for every $p \in V$ and for every $t > 0$,

$$\mu_{Tp}(t) \geq \nu_p\left(\frac{t}{k}\right); \tag{9.1.1}$$

or, equivalently, if there exists a constant $h > 0$ such that, for every $p \in V$ and for every $t > 0$,

$$\mu_{Tp}(ht) \geq \nu_p(t);$$

(d) *strongly **C**-bounded* if there exists a constant $h \in (0,1)$ such that, for every $p \in V$ and for every $t > 0$,

$$\nu_p(t) > 1 - t \Rightarrow \mu_{Tp}(ht) > 1 - ht. \tag{9.1.2}$$

Notice that the definition of a strongly bounded operator in a PN space is naturally suggested by the classical definition of a bounded linear operator; a linear operator T from the normed space $(V, \|\cdot\|)$ into the normed space $(V', \|\cdot\|')$ is said to be bounded if there is a constant $k > 0$ such that, for every $x \in V$,

$$\|Tx\|' \leq k\|x\|. \tag{9.1.3}$$

For this reason these operators were the first to be studied in the context of Šerstnev PN spaces (Prochaska, 1967; Radu, 1975a, 1975b).

As a consequence of (9.1.2), a continuous linear operator on an ordinary normed space is uniformly continuous. The same result holds in PN spaces; this follows immediately from Theorem 9.1.1.

Remark 9.1.1. The identity map I between the PN space (V, ν, τ, τ^*) into itself is *strongly **C**-bounded.* And every linear contracting mappings, according to the definition in Hadzic and Pap (2001) is *strongly **C**-bounded*, i.e. for every $p \in V$ and for every $t > 0$ if the condition $\nu_p(t) > 1 - t$ is satisfied then

$$\nu_{Ip}(ht) = \nu_p(ht) > 1 - ht.$$

And every linear contraction map, according to the definition in the fundamental book by Schweizer and Sklar (1983) is *strongly **B**-bounded.*

Corollary 9.1.1. *If $T : (V_1, \nu, \tau, \tau^*) \to (V_2, \mu, \sigma, \sigma^*)$ is linear and continuous, then it is uniformly continuous.*

The identity map I between any PN space (V, ν, τ, τ^*) and itself is a strongly bounded operator with $k = 1$.

Example 9.1.1. Consider the spaces $C([0,1])$ and $C_1([0,1])$ of the functions that are, respectively, continuous and continuous together with their first derivative on the interval $[0,1]$. They are Banach spaces with respect to the two norms

$$\|f\|_0 := \max_{x\in[0,1]} |f(x)| \quad \text{in } C([0,1]),$$

and

$$\|f\|_1 := \|f\|_0 + \|f'\|_0 \quad \text{in } C_1([0,1]).$$

Choose any d.f. G in Δ^+ different from ε_0 and from ε_∞, and consider the derivative map D from the simple PN space $(C_1([0,1]), \|\cdot\|_1, G, M)$ into the simple PN space $(C([0,1]), \|\cdot\|_0, G, M)$ defined by $Df = f'$. Then, for every $x > 0$, one has

$$\nu'_{Df}(x) = G\left(\frac{x}{\|f\|_0}\right) \geq G\left(\frac{x}{\|f\|_1}\right) = \nu_f(x),$$

whence D is strongly bounded.

Theorem 9.1.2.

(a) *Every strongly bounded operator is certainly bounded;*
(b) *Every strongly bounded operator is bounded.*

Proof. (a) Let T be a strongly bounded operator from (V_1, ν, τ, τ^*) to $(V_2, \mu, \sigma, \sigma^*)$. Then, according to Definition 9.1.1, there exists a constant $k > 0$ such that Eq. (9.1.1) holds for all $p \in V_1$. Then, there exists $t_0 > 0$ such that $\nu_p(t_0) = 1$ for every $p \in A$. Since T is strongly bounded, it suffices to take $t = kt_0$ in Definition 9.1.1 (a) in order to see that $T(A)$ is certainly bounded.
(b) is an immediate consequence of the Definition 9.1.1 (b) and of Eq. (9.1.1). □

The converse of the statements of the above theorem need not be true, as the following example shows.

Example 9.1.2. (A bounded linear operator that is not strongly bounded). Let $V_1 = V_2 = \mathbb{R}, \nu_0 = \mu_0 = \varepsilon_0$, while, if $p \neq 0, \nu_p$ and μ_p are given, respectively, by

$$\nu_p(t) = G\left(\frac{t}{|p|}\right) \quad \text{and} \quad \mu_p(t) = U\left(\frac{t}{|p|}\right)$$

where

$$G(t) := \frac{1}{2} 1_{]0,1[}(t) + 1_{]1,+\infty[}(t),$$

and U is the d.f. of the uniform law on $(0, 1)$

$$U(t) := t 1_{]0,1[}(t) + 1_{]1,+\infty[}(t).$$

Consider now the identity map $I : (\mathbb{R}, |\cdot|, G, M) \to (\mathbb{R}, |\cdot|, U, M)$, by using Example 7.2.3, it is easily proved that I is certainly bounded and bounded, because, for every $k > 0$, for every $p \neq 0$ and for every $t < |p| \min(1/2, k)$,

$$\mu_{Ip}(t) = \mu_p(t) = U\left(\frac{t}{|p|}\right) = \frac{t}{|p|} < \frac{1}{2} = G\left(\frac{t}{k|p|}\right) = \nu_p\left(\frac{t}{k}\right).$$

Moreover, the notions of certain boundedness and boundedness do not imply each other.

Example 9.1.3. Let $(V, \|\cdot\|)$ be a normed space. Let G and G' be in $\Delta^+ \setminus \{\varepsilon_0, \varepsilon_\infty\}$ and consider the identity map I between $(\mathbb{R}, |\cdot|, G, M)$ and $(\mathbb{R}, |\cdot|, U, M)$. Now, with reference to Example 7.3.3:

(a) if $G(t_0)$ for some $t_0 \in]0, +\infty[$, for instance if $G = U$ of the previous example, while $G'(t) < 1$ for every $t \in]0, +\infty[$, but $l^- G'(+\infty) = 1$, for instance, if G' is the d.f. of an exponential law; then I is bounded not certainly bounded;
(b) if $G(t) < 1$ for every $t \in]0, +\infty[$ if $l^- G'(+\infty) = 1$ and if $l^- G'(+\infty) < 1$, then I is certainly bounded but not bounded.

In the classical theory condition (9.1.2) is necessary and sufficient for the continuity of a linear operator. Its analogue in a PN space, namely strong boundedness expressed by (9.1.1), is only sufficient as proved in the following theorem, but not necessary as shown in Example 9.1.4.

Theorem 9.1.3. *Every strongly bounded linear operator T is continuous with respect to the strong topologies in (V_1, ν, τ, τ^*) and $(V_2, \mu, \sigma, \sigma^*)$, respectively.*

Proof. In view of Theorem 9.1.1, it suffices to verify that T is continuous at the origin θ of V. Let $N_{\theta'}(t)$ with $t > 0$ be the arbitrary neighborhood of θ'.

Take $s \leq \min\{t, t/k\}$; then, for every $p \in N_\theta(s)$, one has

$$\mu_{Tp}(t) \geq \nu\left(\frac{t}{k}\right) \geq \nu_p(s) > 1 - s \geq 1 - t.$$

viz. $Tp \in N_{\theta'}$; in other words T is continuous. □

Example 9.1.4. Consider again the simple PN spaces of Example 9.1.2 and the same linear map, i.e., the identity I between them. The map I, which is known not to be strongly bounded, is, however, continuous. Indeed, for every $t > 0$, the neighborhood $N_0(t)$ coincides with the set

$$\left\{p \in \mathbb{R}; |p| < \frac{t}{1-t}\right\}.$$

Since $N_0(s) = \{p \in \mathbb{R} : |p| < s\}$, on taking $s < \min\{\frac{t}{1-t}, \frac{1}{2}\}$, one has, if p is in $N_0(s)$:

$$|p| < s \leq \frac{t}{1-t},$$

so that p belongs also to $N_0(t)$; therefore I is continuous.

The following two examples, together with Example 9.1.2, prove that in the class of linear operators between PN spaces, no two of the concepts of certain boundedness, boundedness and continuity will imply each other.

Example 9.1.5. (A continuous linear operator that is neither certainly bounded nor bounded). Let $(V, \|\cdot\|)$ be a normed space and let F and G be d.f.s in $\mathcal{D}^+$ with $F(t_0) = 1$ for some $t_0 \in]0, +\infty[$. Consider the identity map I from the equilateral space (V, F, Π_M) onto the simple space $(V, \|\cdot\|, G, M)$. Let A be an unbounded set of $(V, \|\cdot\|)$. Then A is certainly bounded in (V, F, Π_M), but it is not $\mathcal{D}$-bounded in $(V, \|\cdot\|, G, M)$. Therefore, I is neither certainly bounded nor bounded.

On the other hand, since the strong topology in an equilateral PM space is discrete (see Section 12.3 in Schweizer and Sklar (1983)) while the strong topology in $(V, \|\cdot\|, G, M)$ is the usual one in $(V, \|\cdot\|)$ since G belongs to $\mathcal{D}^+$, the identity I is continuous.

Example 9.1.6. In the previous example take $F = \varepsilon_a$ with $a \in]0, 1[$: the map I^{-1} from $(V, \|\cdot\|, G, M)$ onto $(V, \varepsilon_a, \Pi_M)$ is certainly bounded and bounded, but it is not continuous, as is immediately seen. In fact, in $(V, \varepsilon_a, \Pi_M)$, any neighborhood $N_\theta(t)$ with $t \in]0, a[$ coincides with the singleton $\{\theta\}$ while any neighborhood $N'_\theta(t)$ with $t > 0$ contains elements of V different from the origin θ.

Theorem 9.1.4. *Let T be a linear map between the two PN spaces (V_1, ν, τ, τ^*) and $(V_2, \mu, \sigma, \sigma^*)$. If there exists a constant $h > 0$ such that, for every $t > 0$ and for every $p \in V$,*

$$\nu_p(t) \geq \mu_{Tp}(hx), \tag{9.1.4}$$

then T has linear inverse T^{-1} that is defined on $T(V)$ and which is strongly bounded.

Proof. Take $T_p = \theta'$ in (9.1.3); then, for every $t > 0, \nu_p(t) \geq 1$, i.e., $\nu_p(t) = 1$, so that $p = \theta$. This yields the existence and the linearity of T^{-1}. Now (9.1.3) can be written in the form

$$\nu_{T^{-1}q}(t) \geq \mu_q(ht),$$

where q is any element of $T(V)$; therefore, T^{-1} is strongly bounded. □

In particular, under the assumptions of the last theorem, the operator T^{-1} is continuous, bounded, and certainly bounded. Moreover, it is hard to check that T maps certainly unbounded sets of (V_1, ν, τ, τ^*) into certainly unbounded sets of $(V_2, \mu, \sigma, \sigma^*)$ and that T maps $\mathcal{D}$-bounded sets of (V_1, ν, τ, τ^*) into $\mathcal{D}$-unbounded sets of $(V_2, \mu, \sigma, \sigma^*)$. The proof of the next two results follow easily from what we have shown.

Corollary 9.1.2. *Let T be a linear map from (V_1, ν, τ, τ^*) into $(V_2, \mu, \sigma, \sigma^*)$ have an inverse T^{-1}. If both the maps T and T^{-1} are strongly bounded then T is a homeomorphism of the PN spaces (V_1, ν, τ, τ^*) and $(V_2, \mu, \sigma, \sigma^*)$.*

The identity of Example 9.1.3 (a) is a homeomorphism and its inverse is not strongly bounded: therefore the converse of Theorem 9.1.4 does not hold in general. The same example shows that the converse of the following corollary may not hold.

Corollary 9.1.3. *Let (V, ν, τ, τ^*) into $(V, \mu, \sigma, \sigma^*)$ be two PN spaces having the same support V. If the identity and its inverse are both strongly bounded, then the strong topologies of the two PN spaces are equivalent.*

In the following example one introduces a strongly **C**-bounded operator, which is not strongly **B**-bounded, nor bounded nor certainly bounded.

Example 9.1.7. Let V be a linear space and let $\nu_\theta = \mu_\theta = \varepsilon_0$, while, if $p, q \neq \theta$ then, for every $p, q \in V$ and $t \in \mathbb{R}_+$, if

$$\nu_p(x) = \begin{cases} 0, & x \leq 1, \\ 1, & x > 1, \end{cases} \quad \mu_p(x) = \begin{cases} \frac{1}{3}, & x \leq 1, \\ \frac{9}{10} & 1 < x < +\infty, \\ 1, & x = +\infty. \end{cases}$$

And if

$$\tau = \tau^* = \sigma = \sigma^* = M,$$

let $I : (V, \nu, \tau, \tau^*) \to (V, \mu, \tau, \tau^*)$ be the identity operator, then I is strongly **C**-bounded but is not strongly **B**-bounded. It is clear that I is not certainly bounded nor bounded.

I is not strongly **B**-bounded, because for every $k > 0$ and for $x > \max\{1, \frac{1}{k}\}$,

$$\mu_{Ip}(kx) = \frac{9}{10} < 1 = \nu_p(x).$$

If one takes any h in (0,1), then for $x > \max\{1, \frac{1}{k}\}$, the condition $\nu_p(x) > 1 - x$ is satisfied and it implies $\mu_{Ip}(hx) = 1 = \nu_p(x) > 1 - hx$.

Definition 9.1.2. Let (V, ν, τ, τ^*) be a PN space. One defines

$$B(p) = \inf\{h \in \mathbb{R} : \nu_p(h^+) > 1 - h\} = d_S(\nu_p, \varepsilon_0) = d_S(\nu_p, \varepsilon_0).$$

Lemma 9.1.1. *Let $T : (V, \nu, \tau, \tau^*) \to (V', \mu, \sigma, \sigma^*)$ be a strongly **B**-bounded linear operator with a constant $k \in (0, 1)$ and let μ_{Tp} be strictly increasing on $[0, 1]$, then for every p in V, $B(Tp) < B(p)$ holds.*

Proof. Let η in $(0, \frac{1-k}{k} B(p))$. Then $B(p) > k[B(p) + \eta]$ and so

$$\eta_{Tp}(B(p)) > \eta_{Tp}(k[B(p) + \eta]).$$

Since T is a strongly **B**-bounded linear operator one has

$$\eta_{Tp}(k[B(p) + \eta]) \geq \nu_p(B(p) + \eta) \geq \nu_p(B(p)^+) > 1 - B(p),$$

and it follows that

$$B(Tp) = \inf\{B(p) : \eta_{Tp}(B(p)^+) > 1 - B(p)\} = d_S(\eta_{Tp}, \varepsilon_0),$$

so that for all p in V one has

$$B(Tp) < B(p).$$

□

Theorem 9.1.5. *Let $T : (V, \nu, \tau, \tau^*) \to (V', \mu, \sigma, \sigma^*)$ be a strongly **B**-bounded linear operator with a constant $k \in (0,1)$ and let μ_{Tp} be strictly increasing on $[0,1]$, then T is a strongly **C**-bounded linear operator.*

Proof. Let T be a strongly **B**-bounded linear operator with constant $k \in (0,1)$. By Lemma 9.1.1, for every $p \in V$ one has $B(Tp) < B(p)$, and there exists $\gamma \in (0,1)$ such that

$$B(Tp) < \gamma B(p).$$

It means that

$$\begin{aligned}\inf\{h \in \mathbb{R} : \eta_{Tp}(h^+) > 1-h\} &\leq \gamma \inf\{h \in \mathbb{R} : \nu_p(h^+) > 1-h\}\\ &= \inf\{\gamma h \in \mathbb{R} : \nu_p(h^+) > 1-h\}\\ &= \inf\left\{h \in \mathbb{R} : \nu_p\left(\frac{h^+}{\gamma}\right) > 1 - \frac{h}{\gamma}\right\}.\end{aligned}$$

We conclude that $\nu_p(\frac{h}{\gamma}) > 1 - \frac{h}{\gamma} \Rightarrow \eta_{Tp}(h) > 1-h$. Now if $x = \frac{h}{\gamma}$, then $\nu_p(x) > 1-x \Rightarrow \eta_{Tp}(x\gamma) > 1-x\gamma$, so T is a strongly **C**-bounded operator. □

Remark 9.1.2. From Theorem 9.1.5 we have noted that under some additional conditions certain types of strongly **B**-bounded operators are also strongly **C**-bounded operators, but not every strongly **B**-bounded operator is strongly **C**-bounded as we can see in the following example.

Example 9.1.8. Let $V = V' = \mathbb{R}$ and $\nu_0 = \mu_0 = \varepsilon_0$ while, if $p \neq 0$, then for $x > 0$, let $\nu_p(x) = G(\frac{x}{|p|})$ and $\mu_p(x) = U(\frac{x}{|p|})$, where

$$G(x) = \begin{cases}\frac{1}{2}, & 0 < x \leq 2,\\ 1, & 2 < x \leq +\infty,\end{cases} \qquad U(x) = \begin{cases}\frac{1}{2}, & 0 < x \leq \frac{3}{2},\\ 1, & \frac{3}{2} < x \leq +\infty.\end{cases}$$

Consider now the identity map $I : (\mathbb{R}, |\cdot|, G, \nu) \to (\mathbb{R}, |\cdot|, U, \mu)$. We can check that I is a strongly **B**-bounded operator but it is not a strongly **C**-bounded.

Theorem 9.1.6. *Let G be a d.f. increasing on $[0,1]$, then $T : (V, \|\cdot\|, G, \alpha) \to (V', \|\cdot\|, G, \alpha)$ is a strongly **B**-bounded operator if, and only if, T is a bounded linear operator between normed spaces.*

Proof. Let $k > 0$ and $x > 0$. Then for every $p \in V$ one has

$$G\left(\frac{kx}{\|Tp\|^{\alpha}}\right) = \mu_{Tp}(kx) \geq \nu_p(x) = G\left(\frac{x}{\|p\|^{\alpha}}\right),$$

if, and only if,

$$\|Tp\| \leq k^{\frac{1}{\alpha}} \|p\|.$$

□

Theorem 9.1.7. *Let $T : (V, \|\cdot\|, G, \alpha) \to (V', \|\cdot\|, G, \alpha)$ be a strongly **C**-bounded operator, and let G be an increasing d.f. on $[0,1]$, then T is a bounded linear operator between normed spaces.*

Proof. If ν_p is increasing for every $p \in V$, then the quasi-inverse $\nu_p^{\wedge}$ is continuous and $B(p)$ is the unique solution of the functional equation $x = \nu_p^{\wedge}(1-x)$, i.e.

$$B(p) = \nu_p^{\wedge}(1 - B(p)). \tag{9.1.5}$$

If $\nu_p(x) = G(\frac{x}{\|p\|^{\alpha}})$, then $\nu_p^{\wedge}(x) = \|p\|^{\alpha} G^{\wedge}(x)$ and from (9.1.5) it follows that

$$B(p) = \|p\|^{\alpha} G^{\wedge}(1 - B(p)). \tag{9.1.6}$$

Suppose that T is strongly **C**-bounded, i.e. that for every $p \in V$, there exists a constant $k \in (0,1)$ such that

$$B(Tp) \leq kB(p), \tag{9.1.7}$$

then (9.1.6) and (9.1.7) imply

$$\|Tp\|^{\alpha} = \frac{B(Tp)}{G^{\wedge}(1 - B(Tp))} \leq \frac{kB(p)}{G^{\wedge}(1 - kB(p))} \leq \frac{kB(p)}{G^{\wedge}(1 - B(p))} = k\|p\|^{\alpha},$$

which means that T is bounded between normed spaces. □

The converse of the above theorem is not true; see Example 9.1.8.

Theorem 9.1.8. *Every strongly **C**-bounded linear operator T is continuous.*

Proof. By Theorem 9.1.1, it suffices to verify that T is continuous at θ. Let $N_{\theta'}(t)$, with $t > 0$, be an arbitrary neighborhood of θ'. There exists a neighborhood of θ, $N_\theta(s)$ such that, for every $p \in N_\theta(s)$ with $s = \frac{t}{h}$ and $h \in (0,1)$, we have for every $t > 0$ that $\nu_p(\frac{t}{h}) > 1 - \frac{t}{h}$. Since T is strongly

C-bounded

$$\mu_{Tp}(hs) \geq 1 - hs;$$

in other words, $Tp \in N_{\theta'}(t)$ and T is continuous. □

9.2 Classes of Linear Operators

Given two PN spaces (V_1, ν, τ, τ^*) into $(V_2, \mu, \sigma, \sigma^*)$. Let $L = L(V_1, V_2)$ be the vector space of linear operators from V_1 to V_2. At the same time, other classes of linear operators will be considered.

- $L_b = L_b(V_1, V_2)$, the subset of L formed by the bounded linear operators from V_1 to V_2;
- $L_c = L_c(V_1, V_2)$, the subset of L formed by the continuous linear operators from V_1 to V_2;
- $L_{bc} = L_{bc}(V_1, V_2)$, the subset of L formed by the continuous and bounded linear operators from V_1 to V_2;

As was shown in the paper by Alsina, Schweizer, and Sklar (1997), PN spaces are not necessarily topological linear spaces. Therefore, that the subsets L_b, L_c, and L_{bc} are linear subspaces of L has to be proved. This is quite easy in the case of L_c, where the usual proof supplemented by the results in Alsina, Schweizer, and Sklar (1997) leads to the result that we state as a theorem.

Theorem 9.2.1. *$L_c = L_c(V_1, V_2)$ is a linear subspace of L.*

However, the sets L_b and L_{bc} are not necessarily linear subspaces of L. A sufficient condition for this is given by the following theorem.

Theorem 9.2.2. *If the triangle function τ_2 maps $\mathcal{D}^+ \times \mathcal{D}^+$ into $\mathcal{D}^+$, i.e., $\sigma(\mathcal{D}^+ \times \mathcal{D}^+) \subset \mathcal{D}^+$, then both $L_b = L_b(V_1, V_2)$ and $L_{bc} = L_{bc}(V_1, V_2)$ are linear subspaces of L.*

Proof. It suffices to show that $L_b = L_b(V_1, V_2)$ is a vector subspace of $L = L(V_1, V_2)$. In this proof A denotes a bounded subset of V_1.

Let T_1 and T_2 be two bounded linear maps from (V_1, ν, τ, τ^*) into $(V_2, \mu, \sigma, \sigma^*)$. Then, by definition of boundedness, both $R'_{T_1(A)}$ and $R'_{T_2(A)}$ are in $\mathcal{D}^+$. Since, for every $p \in A$, one has

$$\nu_{T_1p+T_2p} \geq \sigma(\mu_{T_1p}, \mu_{T_2p}) \geq \sigma(R'_{T_1(A)}, R'_{T_2(A)}),$$

which belongs to $\mathcal{D}^+$, $R'_{(T_1+T_2)A}$ also belongs to $\mathcal{D}^+$ and T_1+T_2 is bounded.

Now let $\alpha \in \mathbb{R}$ and T be in $L_b(V_1, V_2)$. Because of (N2), it suffices to consider the case $\alpha \geq 0$. If either $\alpha = 0$ or $\alpha = 1$, then αT is bounded. Proceeding by induction, assume that αT is bounded, i.e., $R'_{\alpha T(A)} \in \mathcal{D}^+$, for $\alpha = 0, 1, 2, \ldots, n-1$ with $n \in \mathbb{N}$. Then, for every $p \in A$

$$\mu_{nTp} \geq \sigma(\mu_{(n-1)Tp}, \mu_{Tp})$$

and hence

$$R'_{nT(A)} \geq \sigma(R'_{(n-1)T(A)}, R'_{T(A)})$$

so that $R'_{nT(A)}$ belong to $\mathcal{D}^+$ and nT is bounded. Therefore nT is bounded for the positive integer n. If α is not a positive integer, there is $n \in \mathbb{Z}_+$, such that $n - 1 < \alpha < n$. Therefore, by Lemma 3.3.1, for every $p \in A$ one has

$$\mu_{nTp} \leq \mu_{\alpha Tp}$$

whence

$$R'_{nT(A)} \leq R'_{\alpha T(A)}$$

which implies that αT is bounded. □

9.3 Probabilistic Norms for Linear Operators

It is possible to introduce probabilistic norms for linear operators. The following result is crucial for our purposes.

Theorem 9.3.1. *Let A be a subset of V_1 and $\nu^A(T) := R'_{T(A)}$ the probabilistic radius of the image $T(A)$ under T; then the quadruple $(L, \nu^A, \sigma, \sigma^*)$ is a PPN space. Convergence in the probabilistic pseudo norm ν^A is equivalent to uniform convergence of operators on A.*

Proof. For (N1), if Θ is the null operator (i.e., $\Theta_p = \theta_2$ for every $p \in V_1, \theta_2$ being the null vector of V_2), then $R'_{\Theta(A)} = \varepsilon_0$.

Property (N2) is obvious. As for (N3), if S and T belong to L, then, by definition of ν^A, one has

$$\sigma(\nu^A(S), \nu^A(T)) \leq \sigma(\mu_{Sp}, \mu_{Tp}) \leq \mu_{(S+T)p}$$

for every $p \in A$ so that

$$\sigma(\nu^A(S), \nu^A(T)) \leq R'_{(S+T)A} \leq \nu^A(S+T).$$

For (N4), if $\alpha \in [0,1]$ and T is in L, then for every $p \in A$

$$\nu^A(T) = R'_{TA} \leq \mu_{Tp} \leq \sigma^*(\mu_{\alpha Tp}, \mu_{(1-\alpha)Tp}).$$

Therefore, since σ^* is non-decreasing in each variable,

$$\nu^A(T) \leq \sigma^* \left(l^- \inf_{p \in A} \mu_{\alpha Tp}, l^- \inf_{p \in A} \mu_{(1-\alpha)Tp} \right)$$
$$= \sigma^*(\nu^A(\alpha T), \nu^A((1-\alpha)T)).$$

This proves that $(L, \nu^A, \sigma, \sigma^*)$ is a PPN space.

Assume now that $T_n \to T$ as $n \to +\infty$ in the topology of $(L, \nu^A, \sigma, \sigma^*)$; since

$$\nu^A(T_n - T) \leq \mu_{T_n p - Tp}$$

for every $p \in A$, then,

$$d_S(\mu_{T_n p - Tp}, \varepsilon_0) \leq d_S(\nu^A(T_n - T), \varepsilon_0) \quad \text{for every } p \in A,$$

which implies $T_n p \to Tp$ as $n \to +\infty$ uniformly in $p \in A$.

Conversely, assume $T_n p \to Tp$ as $n \to +\infty$ on A; namely for every $\eta > 0$, there exists an $n_0 = n_0(\eta) \in \mathbf{N}$ such that, for every $n \geq n_0$ and for all $p \in A$

$$d_S(\mu_{T_n p - Tp}, \varepsilon_0) < \frac{\eta}{2}$$

or, equivalently,

$$\mu_{T_n p - Tp}\left(\frac{\eta}{2}\right) > 1 - \frac{\eta}{2}.$$

Therefore, for every $n \geq n_0$

$$\nu^A(T_n - T)(\eta) \geq \nu^A(T_n - T)\left(\frac{\eta}{2}\right) \geq 1 - \frac{\eta}{2} > 1 - \eta,$$

viz.

$$d_S(\nu^A(T_n - T), \varepsilon_0) < \eta,$$

which concludes the proof. □

Below we give a condition that ensures that $(L, \nu^A, \sigma, \sigma^*)$ is a PN space. To this purpose we recall that a subset H of a linear space V is said to be a *Hamel basis* if every vector x of V can be expressed in a unique way as a *finite* sum

$$x = \alpha_1 u_1 + \alpha_2 u_2 + \cdots + \alpha_n u_n$$

where, for every $j = 1, \ldots, n$, α_j is a scalar and u_j belongs to H; a subset H of V is a Hamel basis if, and only if, it is a maximal independent set (Dunford & Schwartz, 1957).

Theorem 9.3.2. *If the subset A of V_1 contains a Hamel basis for V_1, then the quadruple $(L, \nu^A, \sigma, \sigma^*)$ is a PN space whose topology is stronger than that of simple convergence for operators, i.e.,*

$$\nu^A(T_n - T) \to \varepsilon_0 \quad as\ n \to +\infty \Longrightarrow \forall p \in V_1 \mu_{T_n p - Tp} \to \varepsilon_0 \quad as\ n \to +\infty.$$

Proof. By Theorem 9.3.1 $(L, \nu^A, \sigma, \sigma^*)$ is a PPN space, and that $\nu^A(T) = \varepsilon_0$ implies $T_p = \theta_2$ for every $p \in A$ (in other words A is contained in the kernel of T, i.e. $A \subset ker(T)$). If p does not belong to A then there exists $n(p) \in \mathbf{N}, \alpha_j \in \mathbb{R}, p_j \in A (j = 1, \ldots, n(p))$ such that

$$p = \Sigma_{j=1}^{n(p)} \alpha_j p_j.$$

Therefore

$$Tp = T(\Sigma_{j=1}^{n(p)} \alpha_j p_j) \Sigma_{j=1}^{n(p)} \alpha_j T p_j = \Sigma_{j=1}^{n(p)} \alpha_j \theta_2 = \theta_2.$$

Thus $Tp = \theta_2$ for every $p \in V_1$, i.e., $T = \Theta$.

If $T_n \to T$ as $n \to +\infty$ in the topology of $(L, \nu^A, \sigma, \sigma^*)$, then as in the proof of Theorem 9.3.1, $T_n p \to Tp$ for every $p \in A$. If p does not belong to A then write $p = \Sigma_{j=1}^{n(p)} \alpha_j p_j$.

Since the operations of vector addition and multiplication by a fixed scalar are continuous in PN space (Alsina, Schweizer & Sklar, 1997), one obtains

$$\begin{aligned} T_n p &= T_n(\Sigma_{j=1}^{n(p)} \alpha_j p_j) \\ &= \Sigma_{j=1}^{n(p)} \alpha_j T_n p_j \to \Sigma_{j=1}^{n(p)} \alpha_j T p_j = T(\Sigma_{j=1}^{n(p)} \alpha_j p_j) = Tp, \end{aligned}$$

which concludes the proof. □

Corollary 9.3.1. *If A is an absorbing subset of V_1, then $(L, \nu^A, \sigma, \sigma^*)$ is a PN space and convergence in the probabilistic norm ν^A is equivalent to uniform convergence of operators on A.*

Proof. As the second statement has the same proof as in Theorem 9.3.1, only the first one will be proved. To this end, we shall show that an absorbing set A contains a Hamel basis for V_1. Let B be a Hamel basis for V_1 and let $p \in B$. Since A is absorbing, there is a scalar $\alpha(p) > 0$ such that $\alpha(p)p \in A$. Then $B' := \{\alpha(p)p; p \in B\}$ is a Hamel basis for V_1. □

The probabilistic norm ν^{V_1} is the analogue of usual operator norm.

Corollary 9.3.2. *The topology of the PN space $(L, \nu^A, \sigma, \sigma^*)$ is equivalent to that of the uniform convergence of operators.*

It ought to be noted that the results we have just presented are stronger than the analogous ones given by Radu (1975) in the special case of those Šerstnev spaces in which $\tau = \tau_T$, in that in the present book the operators of L are only assumed to be linear and not also continuous.

Example 9.3.1. Let F and G be two d.f.s belonging to Δ^+ both different from ε_0 and ε_∞ and such that the relationship $F \leq G$ does not hold. Consider the PN spaces (V_1, G, M) and $(V_2, \nu', \tau_2, \tau_2^*)$, the first of which is equilateral; then consider the equilateral space $(L_{bc}(V_1, V_2), \nu_F, \tau_2, \tau_2^*)$, where, for every $T \in L_{bc}(V_1, V_2)$,

$$\nu_F(T) = \ell^- \inf\{\nu'_{Tp} : \nu_p \geq F\} = \varepsilon_0.$$

Since $\nu_p = G$ for every $p \neq \theta$, $(L_{bc}(V_1, V_2), \nu_F, \tau_2, \tau_2^*)$ is a PN space if, and only if, $L_{bc}(V_1, V_2)$ consists only of the null operator Θ.

In the following we shall consider maps $\psi : \Delta^+ \to \Delta^+$ that satisfy some of the properties:

$$\psi(\varepsilon_0) = \varepsilon_0; \tag{9.3.1}$$

$$\psi(F_1) \leq \psi(F_2) \quad \text{if} \quad F_1 \leq F_2 \quad (F_1, F_2 \in \Delta^+); \tag{9.3.2}$$

if $(V_1, \nu, \tau_1, \tau_1^*)$ and $(V_2, \nu', \tau_2, \tau_2^*)$ are two PN spaces and if T belongs to $L(V_1, V_2)$, then

$$\psi(\nu_p) \leq \nu'_{Tp} \quad \text{for all } p \in V_1; \tag{9.3.3}$$

ψ is continuous in ε_0 with respect to the weak topology, i.e.,

$$d_S(F_n, \varepsilon_0) \to 0 \Rightarrow d_S(\psi(F_n), \psi(\varepsilon_0)) \to 0; \tag{9.3.4}$$

$$\psi(\mathcal{D}^+) \subset \mathcal{D}^+. \tag{9.3.5}$$

Also we shall need the following classes of mappings $\psi : \Delta^+ \to \Delta^+$:

$\Omega_T := \{\psi : \Delta^+ \to \Delta^+\}$ satisfies (9.3.1), (9.3.2), and (9.3.3);
$\Omega_T^c := \{\psi : \Delta^+ \to \Delta^+\}$ satisfies (9.3.1), (9.3.2), (9.3.3), and (9.3.4);
$\Omega_T^b := \{\psi : \Delta^+ \to \Delta^+\}$ satisfies (9.3.1), (9.3.2), (9.3.3), and (9.3.5);
$\Omega_T^{bc} := \{\psi : \Delta^+ \to \Delta^+\}$ satisfies properties (9.3.1) through (9.3.5).

Clearly $\Omega_T^{bc} = \Omega_T^c \cap \Omega_T^b \subset \Omega_T^c \cup \Omega_T^b \subset \Omega_T$.

For $F \in \Delta^+$, let $\sigma(F)$ denote the subset of the PN space $(V_1, \nu, \tau_1, \tau_1^*)$ bounded by F, *viz.*

$$\sigma(F) := \{p \in V_1 : \nu_p \geq F\}.$$

If T in $L(V_1, V_2)$ define $\phi_T : \Delta^+ \to \Delta^+$ via

$$\phi_T(F) := \nu^{\sigma(F)}(T) = R'_{T\sigma(F)}.$$

Starting from the probabilistic pseudonorm introduced in Theorem 9.3.1, in the next two theorems we provide characterizations of the classes of linear operators studied in the previous section.

Theorem 9.3.3. *Let $(V_1, \nu, \tau_1, \tau_1^*)$ and $(V_2, \nu', \tau_2, \tau_2^*)$ be two PN spaces and let T be in $L(V_1, V_2)$. Then*

(a) *ϕ_T belongs to Ω_T;*
(b) *T is in $L_c(V_1, V_2)$ if, and only if, ϕ_T belongs to Ω_T^c;*
(c) *T is in $L_b(V_1, V_2)$ if, and only if, ϕ_T belongs to Ω_T^b;*
(d) *T is in $L_{bc}(V_1, V_2)$ if, and only if, ϕ_T belongs to Ω_T^{bc}.*

Proof.

(a) Property (9.3.1); $\psi(\varepsilon_0) = \nu^{\sigma(\varepsilon_0)}(T) = \nu^{\theta_1}(T) = \varepsilon_0$.
Property (9.3.2); Let $F_1 \leq F_2$. Then $p \in \sigma(F_2)$ implies $\nu_p \geq F_2 \geq F_1$ and hence $p \in \sigma(F_1)$, so that $\sigma(F_2) \subset \sigma(F_1)$. Thus

$$\phi_T(F_2) = \nu^{\sigma(F_2)}(T) = R'_{T\sigma(F_2)} \geq R'_{T\sigma(F_1)} = \nu^{\sigma(F_1)}(T) = \phi_T(F_1).$$

Property (9.3.3); For every $p \in V_1$ one has $p \in \sigma(\nu_p)$, whence, by definition,

$$\phi_T(\nu_p) = \ell^- \inf_{q \in \sigma(\nu_p)} \nu'_{Tq} \leq \nu'_{Tp}.$$

(b) Assume that ϕ_T satisfies (9.3.4) and let $\eta > 0$; then there exists $\delta = \delta(\eta) > 0$ such that $d_S(\phi_T(F), \varepsilon_0) < \eta$ whenever $d_S(F, \varepsilon_0) < \delta$. On the other hand, it follows from (a) that ϕ_T satisfies (9.3.3) so that one has, for every $p \in V_1$,

$$d_S(\nu'_{Tp}, \varepsilon_0) \leq d_S(\phi_T(\nu_p), \varepsilon_0).$$

Therefore, if $d_S(\nu_p, \varepsilon_0) < \delta$ then $d_S(\nu'_{Tp}, \varepsilon_0) < \eta$, in other words, T is continuous.
Conversely, let T be continuous; then, for every $\eta > 0$, there exists $\delta = \delta(\eta) > 0$ such that $d_S(\nu'_{Tp}, \varepsilon_0) \leq \eta/2$ whenever $d_S(\nu_p, \varepsilon_0) < \delta$. Assume

now $F_n \to \varepsilon_0$ in the weak topology, i.e., $d_S(F_n, \varepsilon_0) \to 0$. Because of the definition of $\phi_T(F_n)$, for all $x > 0$ there exists $p_{\eta/2} \in \sigma(F_n)$ such that

$$\phi_T(F_n)(x) \geq \nu'_{Tp_{\eta/2}}(x) - \eta/2. \tag{9.3.6}$$

Since $F_n \to \varepsilon_0$, one has $d_S(F_n, \varepsilon_0) < \delta$ provided n is large enough, say $n \geq n_0$ for a suitable $n_0 = n_0(\delta) \in \mathbb{N}$. Therefore, for every $n \geq n_0$ and for every $p \in \sigma(F_n)$,

$$d_S(\nu_p, \varepsilon_0) \leq d_S(F_n, \varepsilon_0) < \delta,$$

and hence $d_S(\nu'_{Tp}, \varepsilon_0) < \eta/2$. As a consequence, for $n \geq n_0$,

$$\nu'_{Tp}\left(\frac{\eta}{2}\right) > 1 - \frac{\eta}{2}$$

for every $p \in \sigma(F_n)$; in particular, from (9.3.6) one has

$$\phi_T(F_n)(\eta) \geq \phi_T(F_n)\left(\frac{\eta}{2}\right) \geq \nu'_{Tp_{\eta/2}}\left(\frac{\eta}{2}\right) - \frac{\eta}{2} > 1 - \eta,$$

viz. $d_S(\phi_T(F_n), \varepsilon_0) < \eta$ for every $n \geq n_0$.

(c) Let T be bounded and let F be in $\mathcal{D}^+$. Then $\sigma(F)$ is bounded and so is $T\sigma(F)$; therefore $\phi_T(F) = R'_{T\sigma(F)}$ is in $\mathcal{D}^+$. Conversely, if A is a non-empty bounded set of V_1, then R_A belongs to $\mathcal{D}^+$ and $\nu_p \geq R_A$ for every $p \in A$, so that $A \subset \sigma(R_A)$. Therefore $R'_{T(A)} \geq R'_{T\sigma(R_A)} = \phi_T(R_A) \in \mathcal{D}^+$, whence T is bounded.

(d) This now follows from (b) and (c). □

The following result can be proved in a similar manner; therefore its proof will not be given.

Theorem 9.3.4. *Let $(V_1, \nu, \tau_1, \tau_1^*)$ and $(V_2, \nu', \tau_2, \tau_2^*)$ be two PN spaces and let T be in $L(V_1, V_2)$. Then*

(a) *T is in $L_c(V_1, V_2)$ if, and only if, $\Omega_T^c \neq \varnothing$;*
(b) *T is in $L_b(V_1, V_2)$ if, and only if, $\Omega_T^b \neq \varnothing$;*
(c) *T is in $L_{bc}(V_1, V_2)$ if, and only if, $\Omega_T^{bc} \neq \varnothing$.*

Theorem 9.3.5. *If F is in Δ^+ and T is in $L(V_1, V_2)$, then*

(a) *$\phi_T(F) = \max\{\psi(F) : \psi \in \Omega_T\}$;*
(b) *if T is in $L_c(V_1, V_2)$, then $\phi_T(F) = \max\{\psi(F) : \psi \in \Omega_T^c\}$;*
(c) *if T is in $L_b(V_1, V_2)$, then $\phi_T(F) = \max\{\psi(F) : \psi \in \Omega_T^b\}$;*
(d) *if T is in $L_{bc}(V_1, V_2)$, then $\phi_T(F) = \max\{\psi(F) : \psi \in \Omega_T^{bc}\}$.*

Proof. Let T be in $L(V_1, V_2)$ and set $\nu_F(T) := \sup\{\psi(F) : \psi \in \Omega_T\}$. By definition, $\nu_F(T) \geq \psi(F)$ for every $\psi \in \Omega_T$, so that, by Theorem 9.3.3, $\nu_F(T) \geq \phi_T(F)$.

On the other hand one has $\nu'_{Tp} \geq \psi(\nu_p)$ for every $p \in V_1$ and for every $\psi \in \Omega_T$, so that

$$\nu'_{Tp} \geq \psi(\nu_p) \geq \psi(F)$$

for every $p \in \sigma(F)$. Thus one has, for every $p \in \sigma(F)$,

$$\nu'_{Tp} \geq \sup\{\phi(F) : \psi \in \Omega_T\} = \nu_F(T)$$

and hence

$$\phi_T(F) = \ell^- \inf\{\nu'_{Tp} : p \in \sigma(F)\} \geq \nu_F(T).$$

The proof of the remaining assertion is similar. □

Theorem 9.3.6. *Let* $(V_1, \nu, \tau_1, \tau_1^*), (V_2, \nu', \tau_2, \tau_2^*)$, *and* $(V_3, \nu'', \tau_3, \tau_3^*)$ *be three PN spaces and let* T_1 *and* T_2 *be linear operators in* $L(V_1, V_2)$ *and* $L(V_2, V_3)$, *respectively. Then* $T_2 \circ T_1$ *belongs to* $L(V_1, V_3)$ *and*

$$\phi_{T_2 \circ T_1} \geq \phi_{T_1} \circ \phi_{T_2}. \tag{9.3.7}$$

Proof. We need only prove inequality (9.3.7), or, equivalently,

$$R''_{\phi(T_2 \circ T_1)\sigma(F)} \geq R''_{T_2 \sigma(R'_{T_1 \sigma(F)})} \tag{9.3.8}$$

for every $F \in \Delta^+$. Since $A \subset \sigma(R_A)$ for every set A, one has, in particular, $T_1(\sigma(F)) \subset \sigma(R'_{T_1 \sigma(F)})$, which implies

$$(T_2 \circ T_1)\sigma(F) = T_2[T_1(\sigma(F))] \subset T_2 \sigma(R'_{T_1(\sigma(F))}),$$

an inclusion that immediately yields inequality (9.3.8). □

9.4 Completeness Results

It is interesting to study when some of the PN spaces that we have introduced above are complete.

Theorem 9.4.1. *Let* A *be a closed subset of the PN space* $(V_1, \nu, \tau_1, \tau_1^*)$ *that contains a Hamel basis for* V_1. *If the PN space* $(V_2, \nu', \tau_2, \tau_2^*)$ *is complete, then both* $(L(V_1, V_2), \nu^A, \tau_2, \tau_2^*)$ *and* $(L_c(V_1, V_2), \nu^A, \tau_2, \tau_2^*)$ *are complete.*

Proof. Let T_n be a Cauchy sequence in $(L(V_1, V_2), \nu^A, \tau_2, \tau_2^*)$; in other words, for every $\delta > 0$ there exists $n_1 = n_1(\delta) \in \mathbb{N}$ such that for all $n, m \geq n_1$

$$d_S(\nu^A(T_n - T_m), \varepsilon_0) < \delta.$$

Because of the definition of ν^A, one has, for every $p \in A$,

$$d_S(\nu'_{T_n p - T_m p}, \varepsilon_0) \leq d_S(\nu^A(T_n - T_m), \varepsilon_0) < \delta, \tag{9.4.1}$$

so that for every $p \in A$, $T_n p$ is a Cauchy sequence in $(V_2, \nu', \tau_2, \tau_2^*)$, which is complete. Therefore there exists $y_p \in V_2$ such that $T_n p \to y_p$ for every $p \in A$. Since A contains a Hamel basis for V_1, every $p \notin A$ can be represented in the form

$$p = \sum_{i=1}^{n(p)} \alpha_i p_i,$$

where the p_is are in A and belong to a Hamel basis for V_1. Since both addition and product by a fixed scalar are continuous, we can define a linear operator $T : V_1 \to V_2$ through

$$T_p := \begin{cases} y_p, & \text{if } p \in A, \\ \sum_{i=1}^{n(p)} \alpha_i p_i, & \text{if } p \notin A \quad \text{and} \quad p = \sum_{i=1}^{n(p)} \alpha_i p_i. \end{cases}$$

Then $T_n p \to Tp$ uniformly on A, i.e., $T_n \to T$ in the strong topology of the PN space $(L(V_1, V_2), \nu^A, \tau_2, \tau_2^*)$.

In order to show that the PN space $(L_c(V_1, V_2), \nu^A, \tau_2, \tau_2^*)$ is complete it suffices to prove that the limit operator T just obtained is continuous if T_n was a Cauchy sequence in $(L_c(V_1, V_2), \nu^A, \tau_2, \tau_2^*)$.

It follows from the uniform continuity of the probabilistic norm that, for every $\eta > 0$ there exists $\delta = \delta(\eta) > 0$ such that if p, q belong to V_2 and $d_S(\nu'_{p-q}, \varepsilon_0) < \delta$, then $d_S(\nu'_p, \nu'_q) < \eta/2$. Now, since $T_n p$ converges uniformly to Tp, there is $n_0 = n_0(eta) \in \mathbb{N}$ such that $d_S(\nu'_{T_n p - Tp}, \varepsilon_0) < \delta$ for every $p \in V_1$ whenever $n \geq n_0$. Therefore $d_S(\nu'_{T_n p}, \nu'_{Tp}) < \eta/2$ for every $p \in V_1$ when $n \geq n_0$. Since T_{n_0} is continuous, there is $\rho = \rho(\eta) > 0$ such that $d_S(\nu'_{T_{n_0} p}, \varepsilon_0) < \eta/2$ whenever $d_S(\nu_p, \varepsilon_0) < \rho$. Thus

$$d_S(\nu'_{Tp}, \varepsilon_0) \leq d_S(\nu'_{Tp}, \nu'_{T_{n_0} p}) + d_S(\nu'_{T_{n_0} p}, \varepsilon_0) < \eta$$

whenever $d_S(\nu_p, \varepsilon_0) < \rho$, i.e., T is continuous. □

Theorem 9.4.2. *If the PN space $(V_2, \nu', \tau_2, \tau_2^*)$ is complete and if the triangle function τ_2 maps $\mathcal{D}^+ \times \mathcal{D}^+$ into $\mathcal{D}^+$, then the PN spaces $(L_b(V_1, V_2), \nu^{V_1}, \tau_2, \tau_2^*)$ and $(L_{bc}(V_1, V_2), \nu^{V_1}, \tau_2, \tau_2^*)$ are also complete.*

Proof. Let T_n be a Cauchy sequence in $(L_b(V_1, V_2), \nu^{V_1}, \tau_2, \tau_2^*)$; since it is also a Cauchy sequence in $(L(V_1, V_2), \nu^{V_1}, \tau_2, \tau_2^*)$, it converges, by Theorem 9.4.1, to a linear operator T in this latter space. In order to show that T is bounded, let D be a bounded set of V_1, i.e., $R_D \in \mathcal{D}^+$; then one has to prove that there exists a *d.f.* G_D in $\mathcal{D}^+$, such that, for every $p \in D$, $\nu'_{Tp} \geq G_D$. Assume, if possible, that this is not so, namely that there exist $p_0 \in D$ and $\beta < 1$ such that $\nu'_{Tp_0}(x) \leq \beta < 1$ for every $x < 0$. By the same argument as in the previous proof, for every $\eta < (1-\beta)/2$, one has $d_S(\nu'_{T_np}, \nu'_{Tp}) < \eta$ for every $p \in V_1$, whenever $n \geq n_0(\eta)$. For every $x > 0$ there is η small enough to have $x < 1/\eta$; for every such value of η one has, in particular, for every $n \geq n_0$,

$$\nu'_{T_np_0}(x) < \nu'_{Tp_0}(x+\eta) < \beta + \eta < \frac{1+\beta}{2} < 1$$

so that T_nD could not be bounded, a contradiction. As a consequence, TD is bounded. □

9.5 Families of Linear Operators

Definition 9.5.1. A set B of linear operators, $B \subset L(V_1, V_2)$, is said to be *equicontinuous* if, for every $\epsilon > 0$ there exists $\delta = \delta(\epsilon) > 0$ such that, for every $T \in B$ and for every $p \in V_1$, one has

$$d_S(\nu'_{Tp}, \varepsilon_0) < \epsilon \quad \text{whenever} \quad d_S(\nu_p, \varepsilon_0) < \delta.$$

A set B of linear operators, $B \subset L(V_1, V_2)$, is said to be *uniformly bounded* if for every bounded subset A of V_1 there exists a *d.f.* G_A in $\mathcal{D}^+$ such that $R'_{TA} \geq G_A$ for every $T \in B$. In particular, every operator in an equicontinuous family is continuous and every operator in a uniformly bounded family is bounded.

In the following we shall need mappings $\phi : \Delta^+ \to \Delta^+$ that satisfy some of the properties (9.3.1) through (9.3.5) and the other one: if $(V_1, \nu, \tau_1, \tau_1^*)$ and $(V_2, \nu', \tau_2, \tau_2^*)$ are PN spaces and B is a set of linear operators from V_1 into V_2, $B \subset L(V_1, V_2)$, then for all $T \in B$ and for all $p \in V_1$

$$\phi(\nu_p) \leq \nu'_{Tp}. \tag{9.5.1}$$

It is convenient to introduce the following families

$\Omega_B := \{\psi : \Delta^+ \to \Delta^+$ satisfies properties (9.3.1), (9.3.2), and (9.5.1)$\}$;

$\Omega_B^c := \{\psi \in \Omega_B :$ satisfies property (9.3.4)$\}$;

$\Omega_B^b := \{\psi \in \Omega_B :$ satisfies property (9.3.5)$\}$;

$\Omega_B^{bc} := \{\psi \in \Omega_B :$ satisfies properties (9.3.4) and (9.3.5)$\}$.

We can now characterize equicontinuous families and uniformly bounded families of linear operators.

Theorem 9.5.1. *Let $(V_1, \nu, \tau_1, \tau_1^*)$ and $(V_2, \nu', \tau_2, \tau_2^*)$ be two PN spaces, let B be a family of linear operators from V_1 into V_2, $B \subset L(V_1, V_2)$, and define a mapping $\phi_B : \Delta^+ \to \Delta^+$ through*

$$\phi_B(F) := \ell^- \inf\{\nu'_{Tp} : T \in B, p \in \sigma(F)\}.$$

Then

(a) $\phi_B \in \Omega_B$;
(b) *B is equicontinuous if, and only if, ϕ_B belongs to Ω_B^c;*
(c) *B is uniformly bounded if, and only if, ϕ_B belongs to Ω_B^b;*
(d) *B is both equicontinuous and uniformly bounded if, and only if, ϕ_B belongs to Ω_B^{bc}.*

Proof.

(a) is immediate, while the proof of (b) is a simple adaptation of that part (b) of Theorem 9.3.3.
(c) Let $B \subset L(V_1, V_2)$ be uniformly bounded and let F be any *d.f.* in Δ^+. Since $\sigma(F)$ is bounded and hence $R'_{T\sigma(F)} \geq G_{T\sigma(F)}$, this latter being the *d.f.* of Definition 9.5.1, one has $\nu'_{Tp} \geq G_{\sigma(F)}$ which belongs to $\mathcal{D}^+$. Therefore $\phi_B(F) \subset \mathcal{D}^+$.
(d) Proof that if B is both equicontinuous and uniformly bounded then ϕ_B belongs to Ω_B^{bc} is immediate.

Conversely, let A be a bounded subset of V_1 so that R_A is in $\mathcal{D}^+$; since $\nu_p \geq R_A$ for every $p \in A$, one has $A \subset \sigma(R_A)$ so that $R'_{T(A)} \geq \phi_B(R_A) \in \mathcal{D}^+$ for every $T \in B$, whenever B is a uniformly bounded subset of $L(V_1, V_2)$. □

Now one can easily prove the analogues of Theorems 9.3.4 and 9.3.5.

Theorem 9.5.2. *If $(V_1, \nu, \tau_1, \tau_1^*)$ and $(V_2, \nu', \tau_2, \tau_2^*)$ are two PN spaces, and if B is a family of linear operators from V_1 into V_2, $B \subset L(V_1, V_2)$, then*

(a) *B is equicontinuous if, and only if, $\Omega_B^c \neq \emptyset$;*
(b) *B is uniformly bounded if, and only if, $\Omega_B^b \neq \emptyset$;*
(c) *B is both equicontinuous and uniformly bounded if, and only if, $\phi_B^{bc} \neq \emptyset$.*

Theorem 9.5.3. *Let $(V_1, \nu, \tau_1, \tau_1^*)$ and $(V_2, \nu', \tau_2, \tau_2^*)$ be two PN spaces and let B be a family of linear operators from V_1 into V_2, $B \subset L(V_1, V_2)$, then*

(a) $\phi_B = \max\{\phi \in \Omega_B\}$;
(b) $\phi_B = \max\{\phi \in \Omega_B^c\}$, *if B is equicontinuous;*
(c) $\phi_B = \max\{\phi \in \Omega_B^b\}$, *if B is uniformly bounded;*
(d) $\phi_B = \max\{\phi \in \Omega_B^{bc}\}$, *if B is equicontinuous and uniformly bounded.*

Chapter 10

Stability of Some Functional Equations in PN Spaces

Probabilistic functional analysis has emerged as one of the most important mathematical disciplines in view of its necessity in dealing with probabilistic models in applied problems. Probabilistic functional analysis was first initiated by the Prague school of probabilistics led by Spacek and Hans in the 1950s. The stability problem for a functional equation is: assuming that a function approximately satisfies the equation (according to some convention) is it then possible to find "near" this function an exact solution of the considered functional equation. The first stability problem was posed in 1940 by Ulam and partially answered in the affirmative (for Banach spaces) in the next year by Hyers. In this chapter one considers a duality theorem to quickly and easily obtain the unique solution of the functional equation

$$\tau\left(F(j/a), F(j/b)\right) = F(j/a+b) \tag{10.0.1}$$

on the space of probability distribution functions, which in 1978 was first studied and solved by D.H. Mouchtari and A.N. Šerstnev. They showed that if τ is a triangle function then the equality (10.0.1) holds if, and only if, $\tau = \tau_M$. The duality theorem established by M.J. Frank and B. Schweizer (1979) yields a very simple proof of this fact. Our aim is to discuss the stability of the functional equation (10.0.1) by solving for a given ε in $[0,1]$ the inequality

$$d_S(\tau(F(j/a), F(j/b)), F(j/a+b)) \leq \varepsilon, \tag{10.0.2}$$

where F is any distribution in Δ^+, $a, b > 0$, τ is a continuous triangle function, and d_S is the modified Lévy metric.

10.1 Mouchtari–Šerstnev Theorem

Theorem 10.1.1. *The equality* (10.0.1) *holds if, and only if,* $\tau = \tau_M$.

To this end we recall that for any F in Δ^+ the left continuous quasi-inverse of F is the function $F^\wedge$ from $[0,1]$ into $[0,\infty]$ defined by

$$F^\wedge(y) = \begin{cases} 0, & y = 0 \\ \sup\{x | F(x) < y\}, & 0 < y < 1. \end{cases} \qquad (10.1.1)$$

In particular,

$$[F(j/a)]^\wedge = aF^\wedge \qquad (10.1.2)$$

and if $F^\wedge = G^\wedge$ then $F = G$. We denote the space of quasi-inverses of elements of Δ^+ by $(\Delta^+)^\wedge$.

It follows from the duality theorem of M.J. Frank and B. Schweizer (1979) that

$$[\tau_M(F,G)]^\wedge = F^\wedge + G^\wedge, \qquad (10.1.3)$$

whence,

$$\begin{aligned} [\tau_M(F(j/a), F(j/b))]^\wedge &= aF^\wedge + bF^\wedge \\ &= (a+b)F^\wedge \\ &= [F(j/a+b)]. \end{aligned} \qquad (10.1.4)$$

Thus τ_M is a solution of (10.1.1).

Conversely, for any triangle function τ let $\tau^\wedge$ be the binary operation induced on $(\Delta^+)^\wedge$ by

$$\tau^\wedge(F^\wedge, G^\wedge) = [\tau(F,G)]^\wedge. \qquad (10.1.5)$$

Then (10.1.1) is equivalent to

$$\tau^\wedge(aF^\wedge, bF^\wedge) = (a+b)F^\wedge. \qquad (10.1.6)$$

Next, for any F, G in Δ^+ and any $a, b > 0$, let $U^\wedge$ and $V^\wedge$ be the functions defined by

$$U^\wedge = \min\left(\frac{1}{a}F^\wedge, \frac{1}{b}G^\wedge\right) \quad \text{and} \quad V^\wedge = \max\left(\frac{1}{a}F^\wedge, \frac{1}{b}G^\wedge\right). \qquad (10.1.7)$$

Then

$$aU^\wedge \le F^\wedge \le aV^\wedge \quad \text{and} \quad bU^\wedge \le G^\wedge \le bV^\wedge. \qquad (10.1.8)$$

Since $\tau^\wedge$ is non-decreasing on $(\Delta^+)^\wedge$, it follows that

$$\tau^\wedge(aU^\wedge, bU^\wedge) \le \tau^\wedge(F^\wedge, G^\wedge) \le \tau^\wedge(aV^\wedge, bV^\wedge). \qquad (10.1.9)$$

Suppose that τ satisfies (10.1.1). Then combining (10.1.6) and (10.1.10) we have that for all $a, b > 0$,

$$(a+b)U^\wedge \le \tau^\wedge(F^\wedge, G^\wedge) \le (a+b)V^\wedge. \qquad (10.1.10)$$

To show that (10.1.11) implies that $\tau = \tau_M$, we choose x such that $0 < x < 1$ and consider the following three cases:

Case 1. $F^\wedge(x) \neq 0$ and $G^\wedge(x) \neq 0$. Then setting $a = F^\wedge(x)$ and $b = G^\wedge(x)$ in (10.1.8) yields $U^\wedge(x) = V^\wedge(x) = 1$, and using (10.1.11) we have at once that:

$$\tau^\wedge(F^\wedge, G^\wedge)(x) = F^\wedge(x) + G^\wedge(x). \tag{10.1.11}$$

Case 2. $F^\wedge(x) = G^\wedge(x) = 0$. Then setting $a = b = 1$ in (10.1.8) yields $U^\wedge(x) = V^\wedge(x) = 0$, whence by (10.1.11) we have $\tau^\wedge(F^\wedge, G^\wedge)(x) = 0$ and (10.1.12) is again valid.

Case 3. $F^\wedge(x) = 0$ and $G^\wedge(x) \neq 0$. Then setting $a = \varepsilon > 0$ and $b = G^\wedge(x)$ in (10.1.8) yields $U^\wedge(x) = 0$ and $V^\wedge(x) = 1$, whence it follows that $\tau^\wedge(F^\wedge, G^\wedge)(x) \leq G^\wedge(x) + \epsilon$. Since ε is arbitrary, we have $\tau^\wedge(F^\wedge, G^\wedge)(x) \leq G^\wedge(x)$. Therefore from the definition of the triangle function we have $\tau(F, G) \leq G$ whence $G^\wedge(x) \leq \tau^\wedge(F^\wedge G^\wedge)(x)$, again (10.1.12) holds. The same conclusion holds if $F^\wedge(x) \neq 0$ and $G^\wedge(x) = 0$.

Thus (10.1.12) holds for all x in $[0, 1]$ whence, using (10.1.4) and (10.1.6), we have $\tau = \tau_M$.

Remark 10.1.1. In the above arguments, neither the commutativity nor the associativity of τ was used.

Remark 10.1.2. The above arguments shows that $\tau(F(j/a), F(j/b)) \geq F(j/a + b)$ (resp., $\leq F(j/a + b)$) if, and only if, $\tau \leq \tau_M$ (resp., $\tau \geq \tau_M$).

Remark 10.1.3. If L is a suitable binary operation on $[0, \infty]$ then $\tau(F(j/a), F(j/b)) = F(j/a + b)$ if, and only if, $\tau = \tau_{M,L}$ (see Theorem 4.8 in Frank and Schweizer (1979) and Section 7.7 in Schweizer and Sklar (2005).

10.2 Stability of a Functional Equation in PN Spaces

In the set of binary operations on Δ^+ we will consider the metric

$$\hat{d}_S(\tau, \tau') = \sup\{d_S(\tau(F, G), \tau'(F, G)) | F, G \in \Delta^+\}.$$

In order to solve (10.0.2) we need to prove several lemmas. Since τ_M satisfies (10.0.1) then obviously τ_M is a solution of (10.0.2). Now we will see that, in fact, any operation ε- close to τ_M, according to $\hat{d}_S$, is also a solution of (10.0.2).

Lemma 10.2.1. *Let τ be a binary operation on Δ^+ such that $\hat{d}_S(\tau, \tau_M) \leq \varepsilon$ then* (10.2.1) *holds.*

Proof. For any F in Δ^+ and for any $a, b > 0$ we have

$$\begin{aligned}
&d_S(\tau(F(j/a), F(j/b), F(j/a+b))) \\
&\quad \leq d_S(\tau(F(j/a), F(j/b)), \tau_M(F(j/a), F(j/b))) \\
&\qquad + d_S(\tau_M(F(j/a), F(j/b), F(j/a+b))) \\
&\quad = d_S(\tau(F(j/a), F(j/b)), \tau_M(F(j/a), F(j/b))) \\
&\quad \leq \hat{d}_S(\tau, \tau_M) \leq \varepsilon.
\end{aligned}$$

□

The following example shows how to find a large family of operations satisfying the previous lemma.

Example 10.2.1. Let L be a continuous two-place function from $\mathbb{R}^+ \times \mathbb{R}^+$ onto $\mathbb{R}^+$ which is non-decreasing in each place and satisfies

(i) If $x < u$ and $y < v$ then $L(x, y) < L(u, v)$, and
(ii) For all x and y in $\mathbb{R}^+$: $|L(x, y) - x - y|$, then the binary operation $\tau_{M,L}$ defined by

$$\tau_{M,L}(F, G)(x) = \sup\{\min(F(u), G(v)) | L(u, v) = x\}$$

satisfies

$$\hat{d}_S(\tau_{M,L}, \tau_M) \leq \varepsilon. \tag{10.2.1}$$

In order to show (10.2.2) we want to prove for any x in $(0, \frac{1}{\varepsilon})$ the inequalities

$$\tau_{M,L}(F, G)(x) \leq \tau_M(F, G)(x + \varepsilon) + \varepsilon \tag{10.2.2}$$

and

$$\tau_M(F, G)(x) \leq \tau_{M,L}(F, G)(x + \varepsilon) + \varepsilon. \tag{10.2.3}$$

First we note that (10.2.3) is equivalent to prove that for any u_0, v_0 such that $L(u_0, v_0) = x$ is

$$\min(F(u_0), G(v_0)) \leq \sup_{u+v=x+\varepsilon} \{\min(F(u), G(v))\} + \varepsilon.$$

To this end, if $L(u_0, v_0) = x$ and (ii) holds, then we have

$$u_0 + v_0 - \varepsilon \leq L(u_0, v_0) = x.$$

i.e., $u_0 + v_0 \leq x + \varepsilon$. Consider

$$u_1 = u_0 + \frac{x + \varepsilon - u_0 - v_0}{2} \quad \text{and} \quad v_1 = v_0 + \frac{x + \varepsilon - u_0 - v_0}{2}.$$

Then $u_1 \geq u_0, v_1 \geq v_0$ and $u_1 + v_1 = x + \varepsilon$, i.e.,

$$\begin{aligned}\min(F(u_0, G(v_0)) &\leq \min(F(u_1, G(v_1))) \\ &\leq \sup_{u+v=x+\varepsilon} \min(F(u), G(v)) \\ &\leq \sup_{u+v=x+\varepsilon} \min(F(u), G(v)) + \varepsilon\end{aligned}$$

Therefore (10.2.3) follows. Now we proceed to show (10.2.4), i.e., for any $u_2 + v_2 = x$

$$\min(F(u_2), G(v_2)) \leq \sup_{L(u,v)=x+\varepsilon} \min(F(u), G(v)) + \varepsilon.$$

By (ii) again, $L(u_2, v_2) - \varepsilon \leq u_2 + v_2 = x$, i.e., $L(u_2, v_2) \leq x + \varepsilon$. Since the function $f(t) = L(u_2 + t, v_2 + t)$ is a continuous strictly increasing mapping from $\mathbb{R}^+$ onto $\mathbb{R}^+$ and $f(0) = L(u_2, v_2) \leq x + \varepsilon$, there will exist t_0 in $\mathbb{R}^+$ such that $f(t) = L(u_2 + t_0, v_2 + t_0) + x = \varepsilon$. Then,

$$\begin{aligned}\min(F(u_2), G(v_2) &\leq \min(F(u_2 + t_0), G(v_2 + t_0) \\ &\leq \sup_{L(u,v)=x+\varepsilon} \min(F(u), G(v)) \\ &\leq \sup_{L(u,v)=x+\varepsilon} \min(F(u), G(v)) + \varepsilon.\end{aligned}$$

Now we will prove a crucial result in this section.

Let $\Delta_{ic}^+ = \{F | F \in \Delta^+,\ F \text{ is strictly increasing and continuous on } (0, \infty)\}$. It is a well-known fact that Δ_{ic}^+ is dense in Δ^+ with respect to the metric topology defined by d_S.

Lemma 10.2.2. *If a binary operation on Δ^+ which is non-decreasing in each place satisfies* (10.9.1), *then*

$$d_S(\tau(F, G), \tau_M(F, G)) \leq \varepsilon \tag{10.2.4}$$

for all F and G in Δ_{ic}^+.

Proof. Assume that a non-decreasing binary operation τ on Δ^+ satisfies

$$d_S(\tau(H(j/a), H(j/b)), H(j/a + b)) \leq \varepsilon$$

for all $a, b > 0$ and for all H in Δ^+; i.e., for any x in $(0, \frac{1}{\varepsilon})$ we have

$$\tau(H(j/a), H(j/b))(x) \leq H(x + \varepsilon/a + b) + \varepsilon \tag{10.2.5}$$

and

$$H(x/a + b) \leq \tau(H(j/a), H(j/b))(x + \varepsilon) + \varepsilon. \tag{10.2.6}$$

Our aim here is to show that given any couple of functions F and G in Δ^+_{ic} we have

$$d_S(\tau(F, G), \tau_M(F, G)) \leq \varepsilon$$

or, equivalently, that for such F, G in Δ^+_{ic} and for x in $(0, \frac{1}{\varepsilon})$,

$$\tau(F, G)(x) \leq \tau_M(F, G)(x + \varepsilon) + \varepsilon \tag{10.2.7}$$

and

$$\tau_M(F, G)(x) \leq \tau(F, G)(x + \varepsilon) + \varepsilon. \tag{10.2.8}$$

To this end we define f from $[0, 1]$ into $\mathbb{R}$ by

$$f(\lambda) = F(\lambda(x + \varepsilon)) - G((1 - \lambda)(x + \varepsilon)).$$

Since F and G are continuous, f is also continuous, and since F and G are strictly increasing on $(0, \infty)$ we have $f(0) = -G(x + \varepsilon) < 0$ and $f(1) = f(x + \varepsilon) > 0$. Thus there exists $\lambda_0 = 0$, i.e.,

$$F(\lambda_0(x + \varepsilon)) = G((1 - \lambda_0)(x + \varepsilon). \tag{10.2.9}$$

Let U be the distribution function in Δ^+_{ic} defined by

$$U(t) = \max\left(F\left(\frac{\lambda_0 t}{1 - \lambda_0}\right), G(t)\right).$$

The function U satisfies

$$U \geq G \quad \text{and} \quad U\left(\frac{1 - \lambda_0}{\lambda_0} j\right) \geq F, \tag{10.2.10}$$

and by (10.2.10) we also have:

$$U\left(\frac{x + \varepsilon}{\frac{\lambda_0}{1 - \lambda_0} + 1}\right) = U((1 - \lambda_0)(x + \varepsilon)) = F(\lambda_0(x + \varepsilon)) = F((1 - \lambda_0)(x + \varepsilon)). \tag{10.2.11}$$

Using (10.2.6), (10.2.11) and (10.2.12) we obtain the following inequalities:

$$\begin{aligned}
\tau(F,G)(x) &\leq \tau\left(U\left(\frac{1-\lambda_0}{\lambda_0}j\right), U(j/1)\right)(x) \\
&\leq U\left(\frac{x+\varepsilon}{\frac{\lambda_0}{1-\lambda_0}+1}\right)+\varepsilon \\
&= \min(F(\lambda_0(x=\varepsilon)), G((1-\lambda_0)(x+\varepsilon)))+\varepsilon \\
&\leq \sup_{u+v=x+\varepsilon} \min(F(u),G(v))+\varepsilon \\
&= \tau_M(F,G)(x+\varepsilon)+\varepsilon,
\end{aligned}$$

i.e., (10.2.8) holds.

Now let g be the function from $[0,1]$ into $\mathbb{R}$ defined by

$$g(u) = F(\mu x) - G((1-\mu)x).$$

Since F and G are in Δ_{ic}^+, g is continuous, $g(0) = -G(x) < 0 < F(x) = g(1)$, and there exists μ_0 in $(0,1)$ with $g(\mu_0) = 0$, i.e.,

$$F(\mu_0 x) = G((1-\mu_0)x).$$

We will show that

$$\sup_{u+v=x} \min(F(u),G(v)) = F(\mu_0 x) = G((1-\mu_0)x). \tag{10.2.12}$$

This is equivalent to proving that

$$\min(F(u),G(v)) \leq F(\mu_0 x) = G((1-\mu_0)x)$$

for all $u, v > 0$ such that $u + v = x$. If there would exist $u_1, v_1 > 0$ such that $u_1 + v_1 = x$, but

$$\min(F(u_1),G(v_1)) > F(\mu_0 x) = G((1-\mu_0)x),$$

this would imply

$$F(u_1) > F(\mu x) \quad \text{and} \quad G(v_1) > G((1-\mu_0)x).$$

Since F and G are strictly increasing on $(0,1)$, the above inequalities would yield $u_1 > \mu_0 x$ and $v_1 > (1-\mu_0)x$ from this $x = u_1+v_1 > u_0x+(1-\mu_0)x = x$ which is a contradiction. Thus (10.2.13) holds.

Next define the distribution function

$$V(t) = \min\left(F\left(\frac{\mu_0}{1-\mu_0}t\right), G(t)\right).$$

Then,

$$V \leq G \quad \text{and} \quad V\left(\frac{\mu_0}{1-\mu_0}j\right) \leq F. \tag{10.2.13}$$

Moreover, by (10.2.13) we have

$$\begin{aligned}\tau_M(F,G)(x) &= \sup_{u+v=x} \min(F(u),G(v)) \\ &= F(\mu_0 x) = G((1-\mu_0)x) = V((1-\mu_0)x). \end{aligned} \tag{10.2.14}$$

Finally we have (10.2.7), (10.2.13), and (10.2.15):

$$\begin{aligned}\tau_M(F,G)(x) &= V((1-\mu_0)x) = V\left(\frac{x}{\frac{\mu_0}{1-\mu_0}+1}\right) \\ &\leq \tau\left(V\left(\frac{j}{\frac{\mu_0}{1-\mu_0}}\right), V(j/1)\right)(x+\varepsilon)+\varepsilon \\ &= \tau\left(V\left(\frac{1-\mu_0}{\mu_0}j\right), V\right)(x+\varepsilon)+\varepsilon \\ &\leq \tau(F,G)(x+\varepsilon)+\varepsilon\end{aligned}$$

whence (10.2.9) holds. □

Now we will extend (10.2.5) to Δ^+ in the following Lemma.

Lemma 10.2.3. *Let τ be a continuous binary operation on Δ^+ such that*

$$d_S(\tau(F,G),\tau_M(F,G)) \leq \varepsilon \tag{10.2.15}$$

for all F and G Δ^+_{ic}. Then (10.2.15) holds for all F and G in Δ^+ and consequently $\hat{d}_L(\tau_M \leq \varepsilon)$ If a binary operation on Δ^+ which is non-decreasing in each place satisfies (10.9.1), then

$$d_S(\tau(F,G),\tau_M(F,G)) \leq \varepsilon \tag{10.2.16}$$

for all F and G in Δ^+_{ic}. Then (10.2.16) holds for all F and G in Δ_+. If a binary operation on Δ^+ such that

$$d_S(\tau(F,G),\tau_M(F,G)) \leq \varepsilon. \tag{10.2.17}$$

Then (10.2.16) holds for all F and G in Δ_+ and consequently $\hat{d}_S(\tau,\tau_M) \leq \epsilon$.

Proof. Since Δ_{ic}^+ is dense in Δ^+, there exist sequences $\{F_n\}$ and $\{G_n\}$ in Δ_{ic}^+ such that $F = \lim_{n\to\infty} F_n$ and $G = w - \lim_{n\to\infty} G_n$. Since τ and τ_M are continuous and τ satisfies (10.2.16) we will have

$$\lim_{n\to\infty} d_S(\tau(F_n, G_n), \tau(F, G)) = 0, \tag{10.2.18}$$

$$\lim_{n\to\infty} d_S(\tau_M(F_n, G_n), \tau_M(F, G)) = 0 \tag{10.2.19}$$

and

$$d_S(\tau(F_n, G_n), \tau_M(F_n, G_n)) \leq \varepsilon. \tag{10.2.20}$$

Since d_S is a metric we also have by (10.2.18):

$$\begin{aligned} d_S(\tau(F,G), \tau_M(F,G)) &\leq d_S(\tau(F,G), \tau(F_n, G_n)) \\ &\quad + d_S(\tau(F_n, G_n), \tau_M(F_n, G_n)) \\ &\quad + d_S(\tau_M(F_n, G_n), \tau_M(F, G)) \\ &\leq \varepsilon + d_S(\tau(F_n, G_n), \tau_M(F_n, G_n)) \\ &\quad + d_S(\tau_M(F_n, G_n), \tau_M(F, G)), \end{aligned}$$

whence by (10.2.15) and (10.2.16) it follows that for F, G in Δ^+

$$d_L(\tau(F,G), \tau_M(F,G)) \leq \varepsilon.$$

Thus $\hat{d}_L(\tau, \tau_M) \leq \varepsilon$. □

All the previous lemmas yield the general solution of our problem:

Theorem 10.2.1. *Let τ be a continuous non decreasing binary operation on Δ^+ and let $\varepsilon > 0$ be given. Then,*

$$d_S(\tau(F(\jmath/a), F(\jmath/b)), F(\jmath/a+b)) \leq \varepsilon,$$

where F is any distribution in $\Delta^+, a, b > 0$, if, and only if,

$$\hat{d}_S(\tau, \tau_M) \leq \varepsilon.$$

10.3 The Additive Cauchy Functional Equation in RN Spaces: Stability

We recall some useful notions and results. First we evoke the fixed point alternative of Díaz and Margolis, to which we will refer to as:

Lemma 10.3.1. *Let (X, d) be a complete generalized metric space and $J : X \to X$ be a strictly contractive mapping; that is, for every $x, y \in X$*

$$d(Jx, Jy) \leq L d(x, y)$$

for some $L < 1$. *Then, for each fixed element* $x \in X$, *either for every* $n \geq 0$,

$$d(J^n x, J^{n+1} x) = +\infty$$

or for all $n \geq n_0$

$$d(J^n x, J^{n+1} x) < +\infty$$

for some natural number n_0. *Moreover, if the second alternative holds then:*

(i) *the sequence* $(J^n x)$ *is convergent to a fixed point* y^* *of* J;
(ii) y^* *is the unique fixed point of* J *in the set* $Y := \{y \in X, d(J^{n_0} x, y) < +\infty\}$ *and, for all* $x, y \in Y$, $d(y, y^*) \leq \frac{1}{1-L} d(y, Jy)$.

Definition 10.3.1. Let X be a real vector space, F be a mapping from X into $\mathcal{D}^+$ (for any $x \in X$, $F(x)$ is denoted by F_x) and T be a t-norm. The triple (X, F, T) is called a random normed space (RN-space) if, and only if, the following conditions are satisfied:

(RN1) $F_x = \varepsilon_0$ iff $x = \theta$, the null vector,
(RN2) $F_{\alpha x}(t) = F_x(\frac{t}{|\alpha|})$ for all $\alpha \in \mathbb{R} \setminus 0$, and $x \in X$,
(RN3) $F_{x+y}(t_1 + t_2) \geq T(F_x(t_1), F_y(t_2))$, for all $x, y \in X$ and $t_1, t_2 > 0$.

Every normed space $(X, \|\cdot\|)$ defines an RN space (X, F, M) where for every $t > 0$

$$F_u(t) = \frac{t}{t + \|u\|},$$

and M is the minimum t-norm. This space is called the induced *random normed space.*

If the t-norm T is such that $\sup_{0<a<1} T(a, a) = 1$, then every RN space (X, F, T) is a metrizable linear topological space with so called F-topology or (ε, λ)-topology induced by the base of neighborhoods of θ denoted by $\{U(\varepsilon, \lambda) | \varepsilon > 0, \lambda \in (0, 1)\}$, where

$$U(\varepsilon, \lambda) = \{x \in X | F_x(\varepsilon) > 1 - \lambda\}.$$

The following stability results are due to Mihet and Radu (2008).

Let X be a linear space, (Y, F, M) be a complete RN space, and G be a mapping from $X \times \mathbb{R}$ into [0,1], such that $G(x, \cdot) \in \mathcal{D}^+$ for all x in X. Consider the set $E := \{g : X \to Y, g(0) = 0\}$ and the mapping d_G defined on $E \times E$ by

$$d_G(g, h) = \inf\{a \in \mathbb{R}_+, F_{g(x)-h(x)}(at) \geq G(x, t) \text{ for all } x \in X \text{ and } t > 0\},$$

where, as usual, $\inf \varnothing = +\infty$.

Definition 10.3.2. (Šerstnev, 1963) Let (X, μ, T) be an RN space

(1) A sequence $\{x_n\}$ in x is said to be convergent to $x \in X$ if, for every $\varepsilon > 0$ and $\lambda > 0$, there exists a positive integer N such that $\mu_{x_n - x}(\varepsilon) > 1 - \lambda$ whenever $n \geq N$.
(2) A sequence $\{x_n\}$ in x is called a Cauchy if, for every $\varepsilon > 0$ and $\lambda > 0$, there exists a positive integer N such that $\mu_{x_n - x_m}(\varepsilon) > 1 - \lambda$ whenever $n, m \geq N$.
(3) An RN space (X, μ, T) is said to be complete if, and only if, every Cauchy sequence in X is convergent to a point in X.

Theorem 10.3.1. (Schweizer and Sklar, 1983) *If (X, μ, T) is a RN-space and $\{x_n\}$ is a sequence such that $\{x_n\} \to x$, then* $\lim_{n\to\infty} \mu_{x_n}(t) = \mu_x(t)$ *almost everywhere.*

Lemma 10.3.2. *d_G is a complete generalized metric on E.*

Proof. It is immediate that d_G is symmetric and $d_G(f, f) = 0$ for all $f \in E$. If $d_G(f, g) = 0$, then for every fixed x and $t > 0$ one has $F_{f(x)-g(x)}(t) \geq G(x, \frac{t}{a})$ for all $a > 0$. Therefore, $F_{f(x)-g(x)}(t) = 1$ for all x and t, which implies $f = g$ for every $x \in X$. Next, if $d_G(f, g) = a < \infty$ and $d_G(g, h) = b < \infty$, then $F_{f(x)-g(x)}(at) \geq G(x, t)$ and $F_{g(x)-h(x)}(bt) \geq G(x, t)$ for all x and t, therefore $F_{f(x)-h(x)}((a+b)t) \geq Min\{F_{f(x)-g(x)}(at), F_{g(x)-h(x)}(bt)\} \geq G(x, t)$ for all x and t, which shows that $d_G(f, h) \leq a + b$, so that $d_G(f, h) \leq d_G(f, g) + d_G(g, h)$.

Suppose that (g_n) is d_G-Cauchy. We fix x in X and denote $G(x, t)$ by $H(t)$. Let $\varepsilon > 0$ and $\lambda \in (0, 1)$ be given and let $t > 0$ be such that $H(t) > 1 - \lambda$. For $a < \frac{\varepsilon}{t}$, we choose n_0 such that $d_G(g_n, g_m) < a$ for all $n \geq n_0$. Then for all $n \geq n_0$ one has

$$F_{g_n(x)-g_m(x)}(\varepsilon) \geq F_{g_n(x)-g_m(x)}(at) \geq H(t) > 1 - \lambda,$$

hence $(g_n(x))$ is a Cauchy sequence. Since (Y, F, M) is complete, there exists a mapping $g : X \to Y$ with $g(0) = 0$, such that $(g_n(x))$ converges to $g(x)$.

Let $a, \delta > 0$ be given. Then there exists n_0 such that for all $n > n_0$, all $m \geq 1$ and each t one has

$$F_{g_n(x)-g_{(n+m)(x)}}(at) \geq H(t).$$

Fix $n > n_0$ and $t > 0$. Since

$$\begin{aligned} F_{g_n(x)-g(x)}((a+\delta)t) &\geq Min\{F_{g_n(x)-g_{(n+m)(x)}}(at), F_{g_{(n+m)(x)-g(x)}}(\delta t)\} \\ &\geq Min\{H(t), F_{g_{n+m}(x)-g(x)}(\delta t)\}, \end{aligned}$$

by letting $m \to \infty$ we obtain $F_{g_n(x)-g(x)}((a+\delta)t) \geq Min\{H(t), 1\} = H(t)$.

Therefore $d_G(g_n, g) \leq a + \delta$ for every $n \geq n_0$, so that (g_n) is d_G-convergent. □

Theorem 10.3.2. *Let X be a real linear space, let f be a mapping from X into a complete RN space (Y, F, M) with $f(0) = 0$ and let $\Phi : X^2 \to \mathcal{D}^+$ be a symmetric mapping with the property*

$$\forall x, y \in X \quad and \quad \forall t > 0, \quad \exists \alpha \in (0, 2) : \Phi(2x, 2y)(\alpha t) \geq \Phi(x, y)(t). \tag{10.3.1}$$

If for every x, y in X

$$F_{f(x+y)-f(x)-f(y)} \geq \Phi(x, y), \tag{10.3.2}$$

then there is a unique additive mapping $g : X \to Y$ such that for every x in X and $t > 0$,

$$F_{g(x)-f(x)}(t) \geq \Phi(x, x)((2 - \alpha)t). \tag{10.3.3}$$

Moreover,

$$g(x) = \lim_{n\to\infty} \frac{f(2^n x)}{2^n}. \tag{10.3.4}$$

Proof. By setting $y = x$ in (10.3.2), we immediately see that $F_{2f(x)-f(2x)} \geq \Phi(x, x)$ for all x, whence for every $x \in X$, for all $t > 0$ we have

$$F_{f(x)-\frac{f(2x)}{2}}(t) \geq \Phi(x, x)(2t).$$

Let $G(x, t) := \Phi(x, x)(2t)$. Consider the set $E := \{g : X \to Y, g(0) = 0\}$ together with the mapping d_G defined on $E \times E$ by

$$d_G(g, h) = \inf\{a \in \mathbb{R}^+, F_{g(x)-h(x)}(at) \geq G(x, t); x \in X, t > 0\}.$$

By Lemma 10.3.2, (E, d_G) is a complete generalized metric space. Now, let us consider the linear mapping

$$J : E \to E, \quad Jg(x) := \frac{1}{2} g(2x).$$

J is a strictly contractive self-mapping of E with the Lipschitz constant $\frac{\alpha}{2}$. Indeed, let g, h in E be given such that $d_G(g, h) < \varepsilon$. Then for every x in

X and $t > 0$ one has

$$F_{g(x)-h(x)}(\varepsilon t) \geq G(x,t),$$

whence for x in X and for $t > 0$ we have

$$F_{Jg(x)-Jh(x)}\left(\frac{\alpha}{2}\varepsilon t\right) = F_{g(2x)-h(2x)}(\alpha\varepsilon t) \geq G(2x, \alpha t).$$

Since $G(2x, \alpha t) \geq G(x,t)$ for all x and t, then $F_{Jg(x)-Jh(x)}((\frac{\alpha}{2}\varepsilon t) \geq G(x,t)$, that is, $d_G(g,h) < \varepsilon \Rightarrow d_G(Jg, Jh) \leq \frac{\alpha}{2}\varepsilon$. This means that

$$d_G(Jg, Jh) \leq \frac{\alpha}{2} d_G(g,h)$$

for all g, h in E.

Next, from $F_{f(x)-2^{-1}f(2x)}(t) \geq G(x,t)$ it follows that $d_G(f, Jf) \leq 1$.

Using the fixed point alternative we deduce the existence of a fixed point of J, that is, the existence of a mapping $g : X \to Y$ such that for every x in X

$$g(2x) = 2g(x)$$

holds. Moreover, $d_G(f,g) \leq \frac{1}{1-L} d(f, Jf)$ implies the inequality $d_G(f,g) \leq \frac{1}{1-\alpha/2}$ from which it immediately follows $F_{g(x)-f(x)}(\frac{2}{2-\alpha}t) \geq G(x,t)$ for all $t > 0$ and x in X (recall that G is left continuous in the second variable). This means that

$$F_{g(x)-f(x)}(t) \geq G\left(x, \frac{2-\alpha}{2}t\right),$$

for every x in X and $t > 0$, whence we obtain the estimation

$$F_{g(x)-f(x)}(t) \geq \Phi(x,x)((2-\alpha)t),$$

for every x in X and $t > 0$. Since for any x and t, $d_G(u,v) < \varepsilon \Rightarrow F_{u(x)-v(x)}(t) \geq G(x, \frac{t}{\varepsilon})$, from $(J^n f, g) \to 0$, it follows for every x in X

$$\lim_{n\to\infty} \frac{f(2^n x)}{2^n} = g(x).$$

The additivity of g can be proven in the standard way. In fact, since M is continuous, then $z \to F_z$ is continuous (see Chapter 12 in Schweizer and Sklar, 1983). Therefore, for almost all t,

$$\begin{aligned} F_{g(x+y)-g(x)-g(y)}(t) &= \lim_{n\to\infty} F_{\frac{f(2^n(x+y))}{2^n} - \frac{f(2^n x)}{2^n} - \frac{f(2^n y)}{2^n}}(t) \\ &= \lim_{n\to\infty} F_{f(2^n(x+y))-f(2^n x)-f(2^n y)}(2^n t) \\ &\geq \lim_{n\to\infty} \Phi(x,y)\left(\left(\frac{2}{\alpha}\right)^n t\right) = 1, \end{aligned}$$

so that $F_{g(x+y)-g(x)-g(y)}(t) = 1$ for all $t > 0$, which implies

$$g(x+y) - g(x) - g(y) = 0.$$

The uniqueness of g follows from the fact that g is the unique fixed point of J with the following property: "there is $C \in (0, \infty)$ such that $F_{g(x)-f(x)}(Ct) \geq G(x,t)$ for all $x \in X$ and $t > 0$." □

Remark 10.3.1. Except for obvious modifications, the above method can be used to prove the following complementary result:

Theorem 10.3.3. *Let X be a real linear space, let f be a mapping from X into a complete RN space (Y, F, M) with $f(0) = 0$ and let $\Phi : X^2 \to \mathcal{D}^+$ be a symmetric mapping with the property*

$$\exists \alpha \in (0,2)/ \quad \forall x, \ \ y \in X, \ \forall t > 0, \quad \Phi(x,y)(t) \geq \Phi(2x, 2y)(\alpha t). \tag{10.3.5}$$

If the control condition (10.3.2) *holds, then there is a unique additive mapping $g : X \to Y$ such that*

$$\forall x \in X, \forall t > 0, \quad F_{g(x)-f(x)}(t) \geq \Phi\left(\frac{x}{2}, \frac{x}{2}\right)\left(\frac{2-\alpha}{2}\right). \tag{10.3.6}$$

Moreover,

$$g(x) = \lim_{n\to\infty} 2^n f\left(\frac{x}{2^n}\right).$$

Example 10.3.1. Let X and Y be normed spaces and (X, F, M) be the induced RN space. If for every $t > 0$,

$$\Phi(x,y)(t) := \frac{t}{t + \varphi(x,y)},$$

then the condition (10.3.1) holds if, and only if, $\varphi(2x, 2y) \leq \alpha\varphi(x,y)$ for all x,y in X, while 10.3.5 is equivalent to $\varphi(x,y) \leq \alpha\varphi(2x,2y)$. We note that $\varphi(x,y) = \|x\|^p + \|y\|^p$ verifies the first condition for $p < 1$ and the second one in case $p > 1$. Since (10.3.2) reduces to

$$\|f(x+y) - f(x) - f(y)\| \leq \varphi(x,y),$$

for every x, y in X, our theorems slightly extend the results of Aoki and Gajda (see also Cădariu and Radu (2004)).

Remark 10.3.2. In the same way we can prove the following stability result for Jensen equations in RN spaces (compare with Theorem 2.1 in Mirmostafaee and Moslehian (2008)).

Theorem 10.3.4. *Let X be a real linear space, let f be a mapping from X into a complete RN space (Y, F, M) with $f(0) = 0$ and let $\Phi : X^2 \to \mathcal{D}^+$ be* *a symmetric mapping with the following property:*

$$\exists \alpha \in (0,2)/ \quad \forall x, \quad y \in X, \quad \forall t > 0, \quad \Phi(2x, 2y)(\alpha t) \geq \Phi(2, y)(t).$$

If

$$F_{2f(\frac{x+y}{2})-f(x)-f(y)}(t) \geq \Phi(x, y)(t),$$

for all x, y in X and $t > 0$, then there is a unique additive mapping $g : X \to Y$ such that for every x in X and for all $t > 0$, one has

$$F_{g(x)-f(x)}(t) \geq \Phi(2x, 0)((2 - \alpha)t).$$

Remark 10.3.3. In this approach, the t-norm M has been used in proving the triangle inequality for d_G. On the other hand, the problem of replacing M by a weaker t-norm is related to a more difficult problem in the theory of fixed points in RN spaces (see Chapter 3 in the book by Hadžić and Pap (2001)).

10.4 Stability in the Quartic Functional Equation in RN Spaces

Mihet, Saadati, and Vaezpour (2011) prove a stability result for the quartic functional equation in random normed spaces (in the sense of Šerstnev) under arbitrary t-norms and provide an application from random normed spaces under the Lukasiewicz t-norm.

10.4.1 *The quartic functional equation in RN spaces: stability*

Definition 10.4.1. Let X, Y be vector spaces. The equation

$$f(2x + y) + f(2x - y) = 4f(x + y) + 4f(x - y) + 24f(x) - 6f(y) \tag{10.4.1}$$

where the unknown is a mapping $f : X \to Y$ is called a quartic functional equation and every solution of the quartic functional equation is said to be a quartic function.

Theorem 10.4.1. *Let X be a real linear space, (Y, μ, T) be a complete RN space, and $F; X \to Y$ be a mapping with $f(0) = 0$ for which there is $\xi : X^2 \to \mathcal{D}^+$ where $\xi(x, y)$ is denoted by $\xi_{x,y}$ with the property:*

$$\mu_{f(2x+y)+f(2x-y)-4f(x+y)-4f(x-y)-24f(x)+6f(y)}(t) \geq \xi_{x,y}(t) \quad (x, y \in X, t > 0). \tag{10.4.2}$$

If

$$\lim_{n\to\infty} T_{i=1}^{\infty}(\xi_{2^{n+i-1}x,0}(2^{4n+2i}t)) = 1 \tag{10.4.3}$$

and

$$\lim_{n\to\infty} \xi_{2^n x,2^n y}(2^{4n}t) = 1 \tag{10.4.4}$$

for every $x, y \in X$ and $t > 0$, then there exists a unique quartic mapping $Q : X \to Y$ such that

$$\mu_{f(x)-Q(x)}(t) \geq T_{i=1}^{\infty}(\xi_{2^{i-1}x,0}(2^{3i+1}t)) \quad (x, y \in X, t > 0). \tag{10.4.5}$$

Proof. Putting $y = 0$ in (10.4.2), we have

$$\mu_{\frac{f(2x)}{2^4}-f(x)}(t) \geq \xi_{x,0}(2^5 t) \geq \xi_{x,0}(2^4 t). \tag{10.4.6}$$

Therefore,

$$\mu_{\frac{f(2^{k+1}x)}{2^{4(k+1)}}-\frac{f(2^4 x)}{2^{4k}}}\left(\frac{t}{2^{4t}}\right) \geq \xi_{2^k x,0}(2^4 t),$$

which implies

$$\mu_{\frac{f(2^{k+1}x)}{2^{4(k+1)}}-\frac{f(2^4 x)}{2^{4k}}}(t) \geq \xi_{2^k x,0}(2^{4(k+1)}t),$$

that is,

$$\mu_{\frac{f(2^{k+1}x)}{2^{4(k+1)}}-\frac{f(2^4 x)}{2^{4k}}}\left(\frac{t}{2^{k+1}}\right) \geq \xi_{2^k x,0}(2^{3(k+1)}t),$$

for every $k \in \mathbb{N}$ and $t > 0$.

As $1 > 1/2 + \cdots + 1/2^n$, by the triangle inequality it follows that:

$$\begin{aligned}\mu_{\frac{f(2x)}{2^4}-f(x)}(t) &\geq T_{k=0}^{n-1}\left(\mu_{\frac{f(2^{k+1}x)}{2^{4(k+1)}}-\frac{f(2^4 x)}{2^{4k}}}\left(\sum_{k=0}^{n-1}\frac{1}{2^{k-1}}t\right)\right)\\ &\geq T_{k=0}^{n-1}(\xi_{2^k x,0}(2^{3ik+1}t))\\ &= T_{i=1}^{n}(\xi_{2^{i-1}x,0}(2^{3i}t)). \quad (x \in X, t > 0)\end{aligned} \tag{10.4.7}$$

In order to prove the convergence of the sequence $(\frac{f(2^n x)}{2^{4n}})$, we replace x with $2^m x$ in (10.4.7) to find that

$$\mu_{\frac{f(2^{n+m}x)}{2^{4(n+m)}}-\frac{f(2^m x)}{2^{4m}}}(t) \geq T_{i=1}^{n}(\xi_{2^{i+m-1}x,0}(2^{3i+4m}t)). \tag{10.4.8}$$

Since the right-hand side of the inequality tends to 1 as m,n tends to infinity, the sequence $(\frac{f(2^n x)}{2^{4(n)}})$ is a Cauchy sequence. Therefore, we may define $Q(x) = \lim_{n\to\infty}(\frac{f(2^n x)}{2^{4(n)}})$ for all $x \in X$.

Replacing x, y with $2^n x$ and $2^n y$ respectively in (10.4.2), it follows that:

$$\mu_{\frac{f(2^{n+1}x+2^n y)}{2^{4n}}+\frac{f(2^{n+1}x-2^n y)}{2^{4n}}-4\frac{f(2^n x+2^n y)}{2^{4n}}-4\frac{f(2^n x-2^n y)}{2^{4n}}-24\frac{f(2^n x)}{2^{4n}}+6\frac{f(2^n y)}{2^{4n}}}(t)$$
$$\geq \xi_{2^n x, 2^n y}(2^{4n}t). \quad (10.4.9)$$

Taking the limit as $n \to \infty$, we find that Q satisfies (10.4.1) for all $x, y, z \in X$, that is, Q is a quartic map.

To prove (10.4.5), take limit as $n \to \infty$ in (10.4.7).

Finally, to prove the uniqueness of the quartic function Q subject to (10.4.5), let us assume that there exists a quartic function Q' which satisfies (10.4.5). Since $Q(2^n x) = 2^{4n}Q(x)$ and $Q'(2^n x) = 2^{4n}Q'(x)$ for all $x \in X$ and $n \in \mathbb{N}$, from (10.4.5) it follows that:

$$\begin{aligned}\mu_{Q(x)-Q'(x)}(t) &= \mu_{Q(2^n x)-Q'(2^n x)}(2^{4n}t)\\ &\geq T(\mu_{Q(2^n x)-f(2^n x)}(2^{4n-1}t), \mu_{f(2^n x)-Q'(2^n x)}(2^{4n-1}t))\\ &\geq T(T_{i=1}^{\infty}(\xi_{2^{n+i-1}x,0}(2^{4n+3i}t)), T_{i=1}^{\infty}(\xi_{2^{n+i-1}x,0}(2^{4n+3i}t)))\end{aligned}$$

for all $x \in X$. By letting $n \to \infty$, we find that $Q = Q'$. □

Example 10.4.1. Let $(X, \|\cdot\|)$ be a Banach algebra and

$$\mu_x(t) = \begin{cases} max\left\{1 - \frac{\|x\|}{t}, 0\right\}, & t > 0, \\ 0, & t \leq 0. \end{cases}$$

For every $x, y \in X$, let

$$\xi_{x,y}(t) = max\left\{1 - \frac{128\|x\| + 128\|y\|}{t}, 0\right\} \quad (t > 0)$$

and $\xi_{x,y}(t) = 0$ if $t \leq 0$. We note that $\xi_{x,y}$ is a distribution function and

$$\lim_{n\to\infty} \xi_{2^n x, 2^n y}(2^{4n}t) = 1$$

for every $x, y \in X$ and $t > 0$.

It is straightforward to show that (X, μ, W) is an RN space (this was essentially proved by Mushtari (1967) and Radu (2003)). Indeed, $\mu_x(t) = 1 \forall t > 0 \Rightarrow \frac{\|x\|}{t} = 0 \forall t > 0 \Rightarrow x = 0$ and obviously $\mu_{\lambda x}(t) = \mu_x(\frac{t}{\lambda}) \forall x \in X$

and $t > 0$. Next, for every $x, y \in X$ and $t > 0$. Next, for every $x, y \in X$ and $s, t > 0$, we have

$$\begin{aligned}\mu_{x+y}(t+s) &= max\left\{1 - \frac{\|x+y\|}{t+s}, 0\right\} = max\left\{1 - \left\|\frac{x+y}{t+s}\right\|, 0\right\} \\ &= max\left\{1 - \left\|\frac{x}{t+s} + \frac{y}{t+s}\right\|, 0\right\} = max\left\{1 - \left\|\frac{x}{t}\right\| - \left\|\frac{y}{s}\right\|, 0\right\} \\ &= W(\mu_x(t), \mu_y(s)).\end{aligned}$$

It is also easy to see that (X, μ, W) is complete, for

$$\mu_{x-y}(t) \leq 1 - \frac{\|x-y\|}{t} \quad (x, y \in X, t > 0)$$

and $(X, \|\cdot\|)$ is complete.

Define $f : X \to X, f(x) = x^4 + \|x\|x_0$, where x_0 is a unit vector in X. A simple computation shows that

$$\begin{aligned}&\|f(2x+y) + f(2x-y) - 4f(x+y) - 4f(x-y) - 24f(x) + 6f(y)\| \\ &\quad = |\|2x+y\| + \|2x-y\| - 4\|x+y\| - 4\|x-y\| - 24\|x\| + 6\|y\|\|\|x_0\| \\ &\quad \leq 128\|x\| + 128\|y\|\end{aligned}$$

for all $x, y \in X$, hence

$$\mu_{f(2x+y)+f(2x-y)-4f(x+y)-4f(x-y)-24f(x)+6f(y)}(t) \geq \xi_{x,y}(t),$$

for all $x, y \in X$ and $t > 0$.

Fix $x \in X$ and $t > 0$. Then

$$\begin{aligned}W_{i=1}^{\infty}(\xi_{2^{n+i-1}x,0}(2^{4n+3i}t)) &= max\left\{\sum_{i=1}^{\infty}(\xi_{2^{n+i-1}x,0}(2^{4n+3i}t) - 1) + 1, 0\right\} \\ &= max\left\{1 - \frac{64\|x\|}{3.2^{3n}t}, 0\right\},\end{aligned}$$

hence $\lim_{n\to\infty} W_{i=1}^{\infty}(\xi_{2^{n+i-1}x,0}(2^{4n+3i}t)) = 1$.

Thus, all the conditions of Theorem 10.4.2 hold.

Since

$$\begin{aligned}W_{i=1}^{\infty}(\xi_{2^{i-1}x,0}(2^{3i+1}t)) &= max\left\{\sum_{i=1}^{\infty}(\xi_{2^{i-1}x,0}(2^{3i+1}t) - 1) + 1, 0\right\} \\ &= max\left\{1 - \frac{32\|x\|}{3t}, 0\right\},\end{aligned}$$

we deduce that $Q(x) = x^4$ is the unique quartic mapping $Q : X \to X$ such that

$$\mu_{f(x)-Q(x)(t)} \geq max\left\{1 - \frac{32\|x\|}{3t}, 0\right\}$$

for all $x \in X$ and $t > 0$.

10.5 A Functional Equation in Menger PN Spaces

Mihet, Saadati, and Vaezpour (2010) apply the fixed point method to investigate the Hyers–Ulam–Rassias stability for the functional equation

$$\sum_{i=1}^{m} f\left(mx_i + \sum_{j=1, j\neq 1}^{m} x_j\right) + f\left(\sum_{i=1}^{m} x_i\right) = 2f\left(\sum_{i=1}^{m} mx_i\right) \quad (m \in \mathbb{N}, m \geq 2), \tag{10.5.1}$$

where f is an unknown mapping from a real linear space to a class of probabilistic normed spaces. As a particular case, we obtain a Hyers–Ulam–Rassias stability result for Eq. (10.5.1) when X is a quasi-normed space and Y is a p-Banach space, similar to that in Eskandani (2008). A mapping $f : X \to Y$ satisfies (10.5.1) if, and only if, f is additive (see, Eskandani (2008)).

The Menger probabilistic φ-normed space has been introduced by Golet (2007).

Definition 10.5.1. Let φ be a function defined on the real field $\mathbb{R}$ into itself, with the following properties:

(a) $\varphi(-t) = \varphi(t)$ for every $t \in \mathbb{R}$;
(b) $\varphi(1) = 1$;
(c) φ is strictly increasing and continuous on $[0, \infty), \varphi(0) = 0$ and $\lim_{\alpha\to\infty} \varphi(\alpha) = \infty$.

Example 10.5.1. The functions satisfying the above definition are:

$$\varphi(t) = |t|; \quad \varphi(t) = |t|^p, p \in (0, \infty); \quad \varphi(t) = \frac{2t^{2n}}{|t| + 1}, n \in \mathbb{N}.$$

Definition 10.5.2. (Golet, 2007) A Menger probabilistic φ-normed space is a triple (X, ν, T), where X is a real vector space, T is a continuous t-norm, and ν is a mapping from X into $\mathcal{D}^+$ such that the following conditions hold:

(PN1) $\nu_x(t) = \epsilon_0(t)$ for all $t > 0$ if and only if $x = 0$;

(PN2) $\nu_{\alpha x}(t) = \nu_x(\frac{t}{\varphi(\alpha)})$ for all $x \in X, \alpha \neq 0$ and $t > 0$;
(PN3) $\nu_{x+y}(t+s) \geq T(\nu_x(t), \nu_y(s))$ for all $x, y \in X$ and $t, s \geq 0$.

From (PN2) it follows that $\nu_{-x}(t) = \nu_x(t)$ $(x \in X, t \geq 0)$.

Lemma 10.5.1. (Luxemburg–Jung theorem (Jung, 1969)) *Let (X, d) be a complete generalized metric space and $A : X \to X$ be a strict contraction with the Lipschitz constant $L \in (0,1)$, such that $d(x_0, A(x_0)) < +\infty$ for some $x_0 \in X$. Then A has a unique fixed point in the set $Y := \{y \in X, d(x_0, y) < \infty\}$ and the sequence $(A^n(x))_{n \in \mathbb{N}}$ converges to the fixed point x^* for every $x \in Y$. Moreover, $d(x_0, A(x_0)) \leq \delta$ implies $d(x^*, x_0) \leq \frac{\delta}{1-L}$.*

Definition 10.5.3. Let X be a linear space, (Y, ν, T_M) be a complete Menger probabilistic φ-normed space, and G be a mapping from $X \times \mathbb{R}$ into $[0,1]$, such that $G(x, .) \in \mathcal{D}^+$ for all x. Consider the set $E := \{g : X \to Y, g(0) = 0\}$ and the mapping d_G defined on $E \times E$ by

$$d_G(g, h) = \inf\{a \in \mathbb{R}^+, \nu_{g(x)-h(x)}(at) \geq G(x, t) \text{ for all } x \in X \text{ and } t > 0\}$$

where, as usual $\inf \phi = +\infty$.

Lemma 10.5.2. (Mihet and Radu, 2008) *d_G is a complete generalized metric on E.*

10.5.1 *Probabilistic stability of the functional equation* (10.5.1)

For convenience, use the following abbreviation for a given mapping $f : X \to Y$:

$$Df(x_1, x_2, \ldots, x_m) = \sum_{i=1}^{m} f\left(mx_i + \sum_{j=1, j\neq 1}^{m} x_j\right) + f\left(\sum_{i=1}^{m} x_i\right) - 2f\left(\sum_{i=1}^{m} mx_i\right)$$

for all $x_j \in X (1 \leq j \leq m)$.

Theorem 10.5.1. *Let X be a linear space, (Y, ν, T_M) be a complete Menger probabilistic φ-normed space, and let $f : X \to Y$ be a Φ-approximate* *solution of Eq.* (10.5.1), *in the sense that*

$$\nu_{Df(x_1,x_2,\ldots,x_m)}(t) \geq \Phi(x_1,x_2,\ldots,x_m)(t), \forall x_1,x_2,\ldots,x_m \in X, \quad (10.5.2)$$

where Φ *is a mapping from* X^m *to* $\mathcal{D}^+$. *If*

$$\begin{aligned} &\exists \alpha \in (0,\varphi(m)) : \forall x_1,x_2,\ldots,x_m \in X, \forall t > 0, \\ &\Phi(mx_1,mx_2,\ldots,mx_m)(\alpha t) \geq \Phi(x_1,x_2,\ldots,x_m)(t) \end{aligned} \quad (10.5.3)$$

and

$$\lim_{n\to\infty} \alpha^n \varphi(\frac{1}{m^n}) = 0 \quad (10.5.4)$$

then there is a unique additive mapping $g : X \to Y$ *such that*

$$\nu_{g(x)-h(x)}(t) \geq \Phi(x,0,\ldots,0)(\varphi(m)-\alpha)(t), \quad \forall x \in X, \forall t > 0. \quad (10.5.5)$$

Moreover,

$$g(x) = \lim_{n\to\infty} \frac{f(m^n x)}{m^n} \quad (x \in X).$$

Proof. By setting $x_1 = x$ and $x_j = 0$, $(2 \leq j \leq m)$ in (10.5.2), we obtain

$$\nu_{g(x)-h(x)}(t) \geq \Phi(x,0,\ldots,0)(t), \quad \forall x \in X,$$

whence

$$\nu_{m^{-1}f(mx)-f(x)}(t) \geq \Phi(x,0,\ldots,0)(\varphi(m)t), \quad \forall x \in X, \ \forall t > 0.$$

Let $G(x,t) := \Phi(x,0,\ldots,0)(\varphi(m)t)$. Consider the set $E := \{g : X \to Y, g(0) = 0\}$ and the mapping d_G defined on $E \times E$. By Lemma 10.3.2, (E, d_G) is a complete generalized metric space. Now, let us consider the linear mapping $J : E \to E, Jg(x); = \frac{1}{m}g(mx)$. It is easy to check that J is a strictly contractive self-mapping of E with the Lipschitz constant $L = \frac{\alpha}{\varphi(m)}$.

Indeed, let $g, h \in E$ be such that $d_G(g,h) < \varepsilon$. Then

$$\nu_{g(x)-h(x)}(\varepsilon t) \geq G(x,t), \quad \forall x \in X, \ \forall t > 0,$$

whence

$$\nu_{Jg(x)-Jh(x)}\left(\frac{\alpha}{\varphi(m)}\varepsilon t\right) = \nu_{g(mx)-h(mx)}(\alpha\varepsilon t) \geq G(mx,\alpha t).$$

for all $x \in X$ and $t > 0$.

Since $G(mx, \alpha t) \geq G(x, t)$ for all $x \in X$ and $t > 0$, then $\nu_{Jg(x)-Jh(x)}(\frac{\alpha}{\varphi(m)}\varepsilon t) \geq G(x,t)$, that is, $d_G(g,h) < \varepsilon \Rightarrow d_G(Jg, Jh) \leq \frac{\alpha}{\varphi(m)}\varepsilon$. This means that

$$d_G(Jg, Jh) \leq \frac{\alpha}{\varphi(m)} d_G(g,h),$$

for all $g, h \in E$. Next, from

$$\nu_{f(x)-m^{-1}f(mx)}(t) \geq G(x,t)$$

it follows that $d_G(f, Jf) \leq 1$. From Lemma 10.3.1 we deduce the existence of a fixed point of J, that is, the existence of a mapping $g : X \to Y$ such that $g(mx) = mg(x)$, for all $x \in X$.

Also, $d_G(f,g) \leq \frac{1}{1-L} d_G(f, Jf)$ implies the inequality $d_G(f,g) \leq \frac{1}{1-\frac{\alpha}{\varphi(m)}}$ from which it immediately follows $\nu_{g(x)-f(x)}(\frac{\varphi(m)}{\varphi(m)-\alpha}t) \geq G(x,t)$ for all $t > 0$ and $x \in X$. This means that

$$\nu_{g(x)-f(x)}(t) \geq G\left(x, \frac{\varphi(m)-\alpha}{\varphi(m)}t\right) \forall x \in X, \quad \forall t > 0,$$

whence we obtain the estimation

$$\nu_{g(x)-f(x)}(t) \geq \Phi(x,0)((\varphi(m)-\alpha)t) \forall x \in X, \quad \forall t > 0.$$

Since for any $x \in X$ and $t > 0$,

$$d_G(u,v) < \varepsilon \Rightarrow \nu_{u(x)-v(x)}(t) \geq G\left(x, \frac{t}{\varepsilon}\right).$$

From $d_G(J^n f, g) \to 0$, it follows that $\lim_{n\to\infty} \frac{f(m^n x)}{m^n} = g(x)$, for any $x \in X$.

The additivity of g can be proven in the standard way (see Mihet, Saadati and Vaezpour (2010, 2011) and Mirmostafee and Moslehian (2008)). In fact, since T_M is a continuous t-norm, then $z \to \nu_x$ is continuous and thus (see Chapter 12 in Schweizer and Sklar (1983, 2005)),

$$\begin{aligned}
\nu_{Dg(x_1,x_2,\ldots,x_m)}(t) &= \lim_{n\to\infty} \nu_{\frac{Df(m^n x_1, m^n x_2, \ldots, m^n x_m)}{m^n}}(t) \\
&= \lim_{n\to\infty} \nu_{Df(m^n x_1, m^n x_2, \ldots, m^n x_m)}\left(\frac{t}{\varphi(\frac{1}{m^n})}\right) \\
&\geq \lim_{n\to\infty} \Phi(x_1, x_2, \ldots, x_m)\left(\frac{t}{\alpha^n \varphi\left(\frac{1}{m^n}\right)}\right) \\
&= 1.
\end{aligned}$$

We infer that

$$\nu_{Dg(x_1,x_2,\ldots,x_m)}(t) = 1,$$

for all $t > 0$, which implies

$$Dg(x_1, x_2, \ldots, x_m) = 0.$$

The uniqueness of g follows from the fact that g is the unique fixed point of J in the set $\{h \in E, g_G(f, h) < \infty\}$, that is, the only one with property

$$\exists C \in (0, \infty) : \nu_{g(x)-f(x)}(Ct) \geq G(x, t) \quad \text{for all } x \in X \quad \text{and} \quad t > 0.$$

The next corollary provides a Hyer–Ulam–Rassias stability result for Eq. (10.5.1), similar to Theorem 2.2 of Eskandani (2008). □

Corollary 10.5.1. *Let X be a real linear space, let f be a mapping from X into a Banach p-normed space $(Y, \|\cdot\|_p)(p \in (0, 1])$, and let $\Psi : X^m \to \mathbb{R}^+$ be a mapping with the property*

$$\exists \alpha \in (0, m^p) : \Psi(mx_1, mx_2, \ldots, mx_m) \leq |\alpha| \Psi(x_1, x_2, \ldots, x_m), \quad \forall x_1, x_2, \ldots, x_m \in X.$$

If

$$\|Df(x_1, x_2, \ldots, x_m)\|_Y \leq \Psi(x_1, x_2, \ldots, x_m), \quad \forall x_1, x_2, \ldots, x_m \in X,$$

then there is a unique additive mapping $g : X \to Y$ such that

$$\|g(x) - f(x)\|_Y \leq \frac{1}{|m|^p - \alpha} \Psi(x, 0, \ldots, 0), \quad \forall x \in X.$$

Moreover,

$$g(x) = \lim_{n \to \infty} \frac{f(m^n x)}{m^n}.$$

Proof. We have that a p-normed space $(Y, \|\cdot\|_p)$ induces a Menger probabilistic φ-normed space (Y, ν, T_M), through

$$\nu_x(t) = \frac{t}{t + \|x\|_p}$$

and $\varphi(t) = |t|^p$. Indeed, (PN1) is obviously verified and (PN2) follows from

$$\nu_{\alpha x}(t) = \frac{t}{t + \|\alpha x\|_p} = \frac{t}{t + |\alpha|^p \|x\|_p} = \frac{t/|\alpha|^p}{t/|\alpha|^p + \|x\|_p} = \nu_x\left(\frac{t}{|\alpha|^p}\right).$$

Finally, if $\frac{t}{t+\|x\|_p} \leq \frac{s}{s+\|y\|_p}$, then the inequality

$$\frac{t+s}{t+s+\|x+y\|_p} \geq \frac{t}{t+\|x\|_p}$$

follows from $t\|x\|^p + s\|x\|^p \geq t\|x\|^p + t\|y\|^p \geq t\|x+y\|^p$.

If we consider the induced Menger probabilistic φ-normed space (Y, ν, T_M) and the mapping Φ on X^m defined by $\Phi(x_1, x_2, \ldots, x_m)(t) = \frac{t}{t+\Psi(x_1,x_2,\ldots,x_m)}$, then the condition

$$\Phi(mx_1, mx_2, \ldots, mx_m)(\alpha t) \geq \Phi(x_1, x_2, \ldots, x_m)(t)$$

is equivalent to

$$\Psi(mx_1, mx_2, \ldots, mx_m) \leq |\alpha|\Psi(x_1, x_2, \ldots, x_m)$$

and the condition

$$\lim_{n\to\infty} \alpha^n \varphi\left(\frac{1}{m^n}\right) = 0 \tag{10.5.6}$$

reduces to

$$\lim_{n\to\infty} \left(\frac{\alpha}{m^p}\right)^n = 0. \tag{10.5.7}$$

Now the conclusion follows from Theorem 10.5.1. □

The next theorem is the improved version of Theorem 10.5.1.

Theorem 10.5.2. *Let X be a real linear space, let f be a mapping from X into a complete Menger probabilistic φ-normed space (Y, ν, T_M), and let $\Phi : X^m \to \mathcal{D}^+$ be a mapping with the properties:*

$$\exists\alpha \in (0, \varphi(m)) : \Phi(mx, 0, \ldots, 0)(\alpha t) \geq \Phi(x, 0, \ldots, 0)(t), \quad \forall x \in X, \ \forall t > 0 \tag{10.5.8}$$

and

$$\lim_{n\to\infty} \Phi(m^n x_1, m^n x_2, \ldots, m^n x_n)\left(\frac{t}{\varphi(\frac{1}{m^n})}\right) = 1 \tag{10.5.9}$$

$\forall x_1, x_2, \ldots, x_m \in X, \forall t > 0.$

If

$$\nu_{Df(x_1,x_2,\ldots,x_m)}(t) \geq \Phi(x_1, x_2, \ldots, x_m)(t) \quad (x_1, x_2, \ldots, x_m \in X, t > 0) \tag{10.5.10}$$

then there is a unique additive mapping $g : X \to Y$ such that

$$\nu_{g(x)-f(x)}(t) \geq \Phi(x, 0, \ldots, 0)((\varphi(m) - \alpha)t) \quad (x \in X, t > 0). \tag{10.5.11}$$

Proof. The outline of the proof is: the condition (10.5.8) implies $G(mx, \alpha t) \geq G(x, t)$ for all $x \in X$ and $t > 0$, from which it follows that J is a strictly contractive self-mapping of E with the Lipschitz constant $L = \frac{\alpha}{\varphi(m)}$, while the condition (10.5.9) ensures the additivity of g. □

Chapter 11

Menger's 2-Probabilistic Normed Spaces

Lael and Nourouzi introduced the concept of Menger's 2-probabilistic normed space in 2009. In 2012, Harikrishnan, Lafuerza-Guillén, and Ravindran developed the idea of accretive operators, convex sets, compactness, $\mathcal{D}$-boundedness, and their various properties in Menger's 2-probabilistic normed space.

Definition 11.0.1. A pair (X, ν) is called a Menger's 2-probabilistic Normed space (Menger's 2-PN space) if X is a real vector space of $dimX > 1$, ν is a mapping from $X \times X$ into $\mathcal{D}$ (for each $x \in X$, the distribution function $\nu(x, y)$ is denoted by $\nu_{x,y}$ and $\nu_{x,y}(t)$ is the value of $\nu_{x,y}$ at $t \in \mathbb{R}$) satisfying the axioms:

(A1) $\nu_{x,y}(0) = 0$ for all $x, y \in X$;
(A2) $\nu_{x,y}(t) = 1$ for all $t > 0$ if, and only if x, y are linearly dependent;
(A3) $\nu_{x,y}(t) = \nu_{y,x}(t)$ for all $x, y \in X$;
(A4) $\nu_{\alpha x,y}(t) = \nu_{x,y}(\frac{t}{|\alpha|})$ for all $\alpha \in \mathbb{R} \setminus \{0\}$ and for all $x, y \in X$;
(A5) $\nu_{x+y,z}(s + t) \geq \min\{\nu_{x,z}(s), \nu_{y,z}(t)\}$ for all $x, y, z \in X$ and $s, t \in \mathbb{R}$.

We call the mapping $(x, y) \to \nu_{x,y}$ a 2-probabilistic norm (2-P norm) on X.

The geometrical meaning of a 2-P norm on

$$X \text{ is } \nu_{x,y}(t) = P\{(x, y) : \|x, y\| < t\},$$

which is the probability of the set of all $(x, y) \in X \times X$ such that the area of the parallelogram spanned by the vectors x and y is less than t.

From the axioms A1 and A2 of the above definition, it is clear that

$$\nu_{x,y}(t) = \varepsilon_0(t) \Leftrightarrow \text{x and y are linearly dependent.}$$

From a probabilistic point of view this means that for every $t > 0$

$$P\{\|x, y\| < t\} = 1 \Leftrightarrow x = \lambda y, \quad \lambda \neq 0.$$

If one of the points x, y is θ then x and y are linearly dependent and $\|x, y\| = 0$.

Example 11.0.1. Let $(X, \|\cdot, \cdot\|)$ be a 2-normed space. Every 2-norm induces a 2-P norm on X as follows:

$$\nu_{x,y}(t) := \begin{cases} 0, & t \leq 0, \\ \dfrac{t}{t + \|x, y\|}, & t > 0. \end{cases}$$

This 2-P norm is called the standard 2-P norm.

Example 11.0.2. Let $(X, \|\cdot, \cdot\|)$ be a 2-normed space. One defines for every $x, y \in X$ and $t \in \mathbb{R}$ the following 2-P norm

$$\nu_{x,y}(t) := \begin{cases} 0, & t \leq \|x, y\|, \\ 1, & t > \|x, y\|. \end{cases}$$

Then (X, ν) is a 2-PN space.

Definition 11.0.2. Let (X, ν) be a Menger's 2-PN space, and (x_n) be a sequence of X. Then the sequence (x_n) is said to be convergent to x if $\lim_{n \to \infty} \nu_{x_n - x, z}(t) = 1$, i.e. for all $z \in X$ and $t > 0$, and $\alpha \in (0, 1)$, $\exists n_0 \in \mathbb{N}$ such that for every $n > n_0$, one has $\nu_{x_n - x, z}(t) > 1 - \alpha$.

Definition 11.0.3. Let (X, ν) be a Menger's 2-PN space. Then a sequence $(x_n) \in X$ is said to be a Cauchy sequence if $\lim_{n \to \infty} \nu_{x_m - x_n, z}(t) = 1$ for all $z \in X, t > 0$ and $m > n$.

Definition 11.0.4. A Menger's 2-PN space is said to be complete if every Cauchy sequence in X is convergent to a point of X.

A complete Menger's 2-PN space is called a Menger's 2-PN Banach space.

Definition 11.0.5. Let (X, ν) be a Menger's 2-PN space, E be a subset of X, then the closure of E is $\overline{E} = \{x \in X : \exists (x_n) \subset E / x_n \to x\}$, i.e. for $e, \in X, \alpha \in (0, 1)$ and $r > 0$, $x \in \overline{E}$: there exists $n_0 \in \mathbb{N}$ such that for every $n > n_0$ one has $\nu_{x - x_n, e}(r) \geq \alpha$.

We say, E is sequentially closed if $E = \overline{E}$.

Definition 11.0.6. Let E be a subset of a real vector space X then E is said to be a convex set if $\lambda x + (1 - \lambda) y \in E$ for all $x, y \in E$ and $0 < \lambda < 1$.

Definition 11.0.7. Let (X, ν) be a Menger's 2-PN space, for $e, x \in X, \alpha \in (0, 1)$ and $r > 0$ we define the locally ball by,

$$B_{e,\alpha}[x, r] = \{y \in X : \nu_{x-y,e}(r) \geq \alpha\}$$

Definition 11.0.8. Let (X, ν) and (Y, ν') be two Menger's 2-PN spaces, a mapping $T : X \to Y$ is said to be sequentially continuous if $x_n \to x$ implies $T(x_n) \to T(x)$.

Definition 11.0.9 (Pourmoslemi and Salimi, 2008). Let X, Y be two real linear spaces of dimension greater than one and let ν be a function defined on the Cartesian product $X \times Y$ into Δ^+ satisfying the following properties:

(MG2P-N1) $\nu_p(0) = 0$ for all $(x, y) = p \in X \times Y$;
(MG2P-N2) $\nu_{x,y}(t) = 1$ for all $t > 0$ if, and only if $\nu_{x,y} = \varepsilon_0$;
(MG2P-N3) $\nu_{x,y}(t) = \nu_{y,x}(t)$ for all $(x, y) \in X \times Y$;
(MG2P-N4) $\nu_{\alpha x,y}(t) = \nu_{x,\alpha y}(t) = \nu_{x,y}(\frac{t}{\|\alpha\|})$ for every $t > 0$, $\alpha \in \mathbb{R} \setminus \{0\}$ and $(x, y) \in X \times Y$;
(MG2P-N5) $\nu_{x+y,z} \geq \min\{\nu_{x,z}, \nu_{y,z}\}$ for every $x, y \in X$ and $z \in Y$;
(MG2P-N6) $\nu_{x,y+z} \geq \min\{\nu_{x,y}, \nu_{x,z}\}$ for every $x \in X$ and $y, z \in Y$.

The function ν is called a Menger generalized 2-probabilistic norm on $X \times Y$ and the pair $(X \times Y, \nu)$ is called a Menger generalized 2-probabilistic normed space (MG2PN space).

Definition 11.0.10. Let $A \times B$ be a non-empty subset of a MG2PN space $(X \times Y, \nu)$ then its probabilistic radius $R_{A\times B}$ is defined by

$$R_{A\times B}(x) := \begin{cases} l^{-}\varphi_{A\times B}(x), & x \in [0, +\infty), \\ 1, & x = \infty. \end{cases}$$

where $\varphi_{A\times B}(x) := \inf\{\nu_{x,y}(x) : x \in A, y \in B\}$.

Definition 11.0.11. Let $A \times B$ be a non-empty subset of a MG2PN space $(X \times Y, \nu)$, then $A \times B$ is said to be:

(1) *certainly bounded*, if $R_{A\times B}(x_0) = 1$ for some $x_0 \in (0, \infty)$;
(2) *perhaps bounded*, if one has $R_{A\times B}(x) < 1$ for every $x \in (0, \infty)$ and $l^{-1}R_{A\times B}(+\infty) = 1$;
(3) *perhaps unbounded*, if $R_{A\times B}(x_0) > 0$ for some $x_0 \in (0, \infty)$ and $l^{-1}R_{A\times B}(+\infty) \in (0, 1)$;
(4) *certainly unbounded*, if $l^{-1}R_{A\times B}(+\infty) = 0$.

A is said to be $\mathcal{D}$-bounded if either (1) or (2) holds.

Theorem 11.0.1 (Lafuerza-Guillén, 2001; Pourmoslemi and Salimi, 2008). *Let $(X \times Y, \nu)$ and $A \times B$ be a Menger's G2PN space and a $\mathcal{D}$-bounded subset of $X \times Y$ respectively. The set $\alpha A \times B := \{(\alpha p, q) : p \in A, q \in B\}$ is also $\mathcal{D}$-bounded for every fixed $\alpha \in \mathbb{R} \setminus \{0\}$ if $\mathcal{D}^+$ is a closed set under the t-norm M.*

11.1 Accretive Operators in 2-PN Spaces

Let (X, ν) be a 2-PN space and $A : D(A) \subset X \to X$ be an operator with domain $D(A) = \{x \in X; Ax \neq 0\}$ and range $R(A) = \cup\{Ax; x \in D(A)\}$. We may identify A with its graph and the closure of A with the closure of its graph.

Definition 11.1.1. Let (X, ν) be a 2-PN space. An operator $A : D(A) \subset X \to X$ is said to be *accretive* if for every $z \in D(A)$

$$\nu_{x-y,z}(t) \geq \nu_{(x-y)+\lambda(Ax-Ay),z}(t) \quad \text{for all } x, y \in D(A) \quad \text{and} \quad \lambda > 0.$$

The notation $[x, y] \in A$ means $x, y \in X$ such that $y = Ax$. Let A be an accretive operator in a 2-PN space (X, ν). Define the resolvent of A by $J_\lambda = (1 + \lambda A)^{-1}$ and the Yosida approximation $A_\lambda = \frac{1}{\lambda}(I - J_\lambda)$ for every $\lambda > 0$. Then $D(J_\lambda) = R(I + \lambda A), R(J_\lambda) = D(A), D(A_\lambda) = D(J_\lambda)$ for $t > 0$. Some properties of J_λ are following:

Lemma 11.1.1. *Let A be an accretive operator in a 2-PN space (X, ν), J_λ is single valued and*

(i) $\nu_{J_\lambda(x)-J_\lambda(y),z}(t) \geq \nu_{x-y,z}(t)$
(ii) $\nu_{\frac{1}{n}[J_\lambda^n(x)-x],z}(t) \geq \nu_{J_\lambda(x)-x,z}(t)$

for all $x, y \in D(J_\lambda), \lambda > 0, z \in X$.

Proof. Let $x, y \in D(J_\lambda), \lambda > 0$ and $t \in \mathbb{R}$.

Suppose $y_1 = J_\lambda(x), y_2 = J_\lambda(x)$. Since A is accretive,

$$\nu_{y_1-y_2,z}(t) \geq \nu_{(y_1-y_2)+\lambda[\frac{1}{\lambda}(x-y_1)-\frac{1}{\lambda}(x-y_2)],z}(t) = \nu_{0,z}(t) = \varepsilon_0(t) \quad \text{for all } z \in X$$

$\Rightarrow y_1 - y_2, z$ are linearly independent for every $z \in X$
$\Rightarrow y_1 - y_2 = 0 \Rightarrow y_1 = y_2$.

Therefore, there exists $[x_1, y_1], [x_2, y_2] \in A$ such that $x_1 + \lambda y_1 = x_2 + \lambda y_2$ implies $J_\lambda(x) = x_1, J_\lambda(y) = x_2$.

(i) Since A is accretive,

$$\begin{aligned}\nu_{J_\lambda(x)-J_\lambda(y),z}(t) = \nu_{x_1-x_2,z}(t) &\geq \nu_{[(x_1-x_2)+\lambda(y_1-y_2)],z}(t)\\ &= \nu_{[(x_1+\lambda y_1)-(x_2+\lambda y_2)],z}(t)\\ &= \nu_{x-y,z}(t), \text{ for every } z \in X.\end{aligned}$$

(ii) We have,

$$\begin{aligned}&\nu_{\frac{1}{n}[J_\lambda^n(x)-x],z}(t) = \nu_{J_\lambda^n(x)-x,z}(nt)\\ &= \nu_{[J_\lambda^n(x)-J_\lambda^{n-1}(x)+J_\lambda^{n-1}(x)-x],z}[t+(n-1)t]\\ &= \min\{\nu_{[J_\lambda^n(x)-J_\lambda^{n-1}(x)],z}(t)\nu_{[J_\lambda^{n-1}(x)-x],z}[(n-1)t]\}\\ &\geq \min\{\nu_{[J_\lambda^n(x)-J_\lambda^{n-1}(x)],z}(t),\\ &\qquad \min\{\nu_{[J_\lambda^{n-1}(x)-J_\lambda^{n-2}(x)],z}(t), \nu_{[J_\lambda^{n-2}(x)-x],z}[(n-2)t]\}\}\\ &\geq \min\{\nu_{[J_\lambda(x)-x,z}(t) \min\{\nu_{[J_\lambda(x)-x],z}(t)\ldots\}\\ &\qquad \min\{\nu_{[J_\lambda(x)-x],z}(t)\ldots\}\}\\ &\geq \nu_{J_\lambda(x)-x,z}(t) \quad \text{for all } z \in X.\end{aligned}$$

□

Definition 11.1.2. Let (X, ν) be a 2-PN space. An operator $A : D(A) \subset X \to X$ is said to be *m-accretive* if $R(I + \lambda A) = X$ for $\lambda > 0$.

Let $A : D(A) \subset X \to X$ and $B : D(B) \subset X \to X$ be two operators. Then B is said to be an *extension* of A if $D(A) \subset D(B)$ and $Ax = Bx$ for every $x \in D(A)$, denote it by $A \subset B$.

Definition 11.1.3. Let (X, ν) be a 2-PN space. An operator $A : D(A) \subset X \to X$ is said to be a *maximal accretive operator* in X if A is an accretive operator in X and for every accretive operator B of X with $A \subset B$ then $A = B$.

Theorem 11.1.1. *Let (X, ν) be a 2-PN space and A be an m-accretive operator in X, then A is a maximal accretive operator.*

Proof. Let B be an accretive operator with $A \subset B$. Let $\lambda > 0$ and $[x, y] \in B$.

Since A is m-accretive we have $x + \lambda y \in R(I + \lambda A)$, which implies there exists $[x_1, y_1] \in A$ such that $x + \lambda y = x_1 + \lambda y_1$.

Since B is accretive and $[x_1, y_1] \in B$ we have,

$$\begin{aligned}\nu_{x-x_1,z}(t) \geq \nu_{(x-x_1)+\lambda(Bx-Bx_1),z}(t) &= \nu_{(x-x_1)+\lambda(y-y_1),z}(t)\\ &= \nu_{(x+\lambda y)-(x_1+\lambda y_1),z}(t)\end{aligned}$$

$$= \nu_{0,z}(t) \quad \text{for every } z \in X$$
$$= \varepsilon_0(t) \quad \text{for every } z \in X$$

$\Rightarrow x - x_1 = 0 \Rightarrow x = x_1$.

Therefore $y = y_1$ implies $[x, y] \in A$. So $A = B$.

Hence A is a maximal accretive operator. □

Lemma 11.1.2. *Let A be an accretive operator in a 2-PN space (X, ν) and let $(u, v) \in X \times X$; then A is maximal accretive in X iff $\nu_{(x-u),z}(t) \geq \nu_{(x-u)+\lambda(y-v),z}(t)$ for every $[x, y] \in A, z \in X$ and $\lambda > 0$ implies $[u, v] \in A$.*

Proof. Let A be a maximal accretive operator in X. Put $T = A \cup [u, v]$. Suppose $\nu_{(x-u),z}(t) \geq \nu_{(x-u)+\lambda(y-v),z}(t)$ for every $[x, y] \in A, z \in X$ and $\lambda > 0$, then T is accretive in X and $A \subset T$ implies $[u, v] \in A$.

Conversely, suppose that if A is an accretive operator in X and $\nu_{(x-u),z}(t) \geq \nu_{(x-u)+\lambda(y-v),z}(t)$ for every $[x, y] \in A, z \in X$ and $\lambda > 0$, which implies $[u, v] \in A$.

Let B be accretive in X with $A \subset B$ and $[x_1, y_1] \in B$. Since B is accretive in X, we have for every $[x, y] \in A, z \in X$ and $\lambda > 0$ with

$$\nu_{x-x_1,z}(t) \geq \nu_{(x-x_1)+\lambda(Bx-Bx_1),z}(t) = \nu_{(x-x_1)+\lambda(y-y_1),z)} \Rightarrow [x_1, y_1] \in A.$$

Therefore $B \subset A$. So $A = B$.

Hence A is maximal accretive in X. □

Theorem 11.1.2. *Let A be an accretive operator in a 2-PN space (X, ν); then the closure $\overline{A}$ of A is accretive.*

Proof. Let $[x_1, y_1], [x_2, y_2] \in \overline{A}$, then there exist sequences $\{[x_n, y_n]\}$, $\{[x_m, y_m]\}$ in A such that $x_n \to x_1; y_n \to y_1; x_m \to x_2; y_m \to y_2$ and $\lambda > 0$.

Since A is accretive we have,

$$\nu_{x_n-x_m,z}(t) \geq \nu_{(x_n-x_m)+\lambda(Ax_n-Ax_m),z}(t) \quad \text{for every } z \in X$$
$$= \nu_{(x_n-x_m)+\lambda(y_n-y_m),z}(t) \quad \text{for every } z \in X$$
$$\text{as } n \to \infty, \nu_{x_1-x_2,z}(t) \geq \nu_{(x_1-x_2)+\lambda(y_1-y_2),z}(t) \quad \text{for every } z \in X$$

$\Rightarrow \overline{A}$ is accretive in X. □

Theorem 11.1.3. *Let A be a maximal accretive operator in a 2-PN space (X, ν) then A is sequentially closed.*

Proof. For all $x_n, y_n \in D(A)$, let $\{[x_n, y_n]\}$ in A such that $x_n \to u, y_n \to v$ and $\lambda > 0$.

Since A is accretive in X and $[x,y] \in A$ implies $\nu_{x-x_n,z}(t) \geq \nu_{(x-x_n)+\lambda(y-y_n),z}(t)$ for every $z \in X$, as $n \to \infty$, we have $\nu_{x-u,z}(t) \geq \nu_{(x-u)+\lambda(y-v),z}(t)$ for every $z \in X$.

Therefore, by Lemma 11.1.2, $[u,v] \in A$. Hence A is sequentially closed. □

Corollary 11.1.1. *If A is an m-maximal accretive operator in a 2-PN space (X, ν) then A is sequentially closed.*

Proof. Since every m-accretive operator A in X is a maximal accretive operator in X, by Theorem 11.1.3, A is sequentially closed. □

Theorem 11.1.4. *Let (X, ν) be a complete 2-PN space. Let A be a sequentially continuous accretive operator on X. If A is closed then $R(I + \lambda A)$ is closed for $\lambda > 0$.*

Proof. Let $\{z_n\}$ be a sequence in $R(I+\lambda A)$ with $z_n \to z'$ in X, then $\{z_n\}$ is a Cauchy sequence in X.

Since $\{z_n\} \in R(I + \lambda A)$, there exists $[x_n, y_n] \in A$ such that $z_n = x_n + \lambda y_n \Rightarrow J_\lambda(z_n) = x_n$.

Therefore for every $t \in R$ and $z \in X$,

$$\nu_{x_n - x_m, z}(t) = \nu_{J_\lambda(z_n) - J_\lambda(z_m), z}(t) \geq \nu_{z_n - z_m, z}(t)$$

$\Rightarrow \underline{\lim}_{n,m\to\infty} \nu_{x_n - x_m, z}(t) = \varepsilon_0(t) = 1$ for every $t > 0$

$\Rightarrow \lim_{n,m\to\infty} \nu_{x_n - x_m, z}(t) = 1$ for every $t > 0$ and $z \in X$.

Therefore $\{x_n\}$ is a Cauchy sequence in X.

Since X is complete, there exists $x \in X$ such that $x_n \to x$ and $y_n = \frac{1}{\lambda}(z_n - x_n)$

$\Rightarrow y_n \to \frac{1}{\lambda}(z' - x)$ as $n \to \infty$.

Since $Ax_n = y_n$ and A is sequentially continuous, $Ax = \frac{1}{\lambda}(z' - x)$

$\Rightarrow z' = x + \lambda Ax \in R(I + \lambda A)$. Hence $R(I + \lambda A)$ is closed for $\lambda > 0$. □

11.2 Convex Sets in 2-PN Spaces

Theorem 11.2.1. *Every open ball in a 2-PN space (X, ν) is convex.*

Proof. A locally ball in 2-PN space is $B_{e,\alpha}[x,r] = \{y \in X : \nu_{x-y,e}(r) \geq \alpha\}$.

Let $x, e \in X$ and $r \in (0,1)$.

Choose $z, y \in$ and $0 \leq \lambda \leq 1$ then $\nu_{x-z,e}(r) \geq \alpha, \nu_{x-y,e}(r) \geq \alpha$.

We have,

$$\begin{aligned}\nu_{x-[\lambda y+(1-\lambda)z],e}(r) &= \nu_{x-[\lambda y+(1-\lambda)z],e}([\lambda+(1-\lambda)]r)\\ &= \nu_{\lambda(x-y)+(1-\lambda)(x-z),e}([\lambda+(1-\lambda)]r)\\ &\geq \min\{\nu_{\lambda(x-y),e}(\lambda r), \nu_{(1-\lambda)(x-z),e}((1-\lambda)r)\}\\ &= \min\{\nu_{(x-y),e}(r), \nu_{(x-z),e}(r)\}\\ &\geq \min\{\alpha,\alpha\} = \alpha.\end{aligned}$$

Therefore $\lambda y+(1-\lambda)z \in B_{e,\alpha}[x,r]$ for all $z, y \in B_{e,\alpha}[x,r]$. Hence, every locally ball in a 2-PN space is convex. □

Theorem 11.2.2. *The closure of a closed convex set in a 2-PN space (X,ν) is convex.*

Proof. Let E be a closed convex set in X. Then we have to prove that $\overline{E}$ is convex. Let $x, y \in \overline{E}$, then there exist sequences $\{x_n\}, \{y_n\} \in E$ such that $x_n \longrightarrow x$ and $y_n \longrightarrow y$. Since, $\{x_n\}, \{y_n\} \in E$ and E is convex implies $\lambda x_n + (1-\lambda)y_n \in E$ for all $0 < \lambda < 1$, as $n \longrightarrow \infty$ we get $\lambda x + (1-\lambda)y \in E$. The facts $\lambda x + (1-\lambda)y \in E$ and $E = \overline{E}$ imply $\overline{E}$ is convex. □

Definition 11.2.1. Let E be a subset of a 2-PN space (X,ν). Then an element $x \in E$ is called an interior point of E if there are $r > 0, e \in X$ such that $B_{e,\alpha}[x,r] \subseteq E$. The set of all interior points of E is denoted by $int(E)$.

Definition 11.2.2. A subset E of a 2-PN space (X,ν) is said to be open if $E = int(E)$.

For any two points x, y in the real vector space X, denote

$$(x,y) = \{\lambda x + (1-\lambda)y; \lambda \in (0,1)\}.$$

Theorem 11.2.3. *Let E be a convex subset of a 2-PN space (X,ν). Let $a \in E$ and if x is an interior point of E then every point in $(a,x) = \{\lambda a + (1-\lambda)x; \lambda \in (0,1)\}$ is an interior point of E.*

Proof. Let $u \in (a,x)$ then $u = \lambda x + (1-\lambda)a$ for some $\lambda \in (0,1)$.

Since x is an interior point of E, then there exists $r_0 > 0, e \in X$ and $\alpha \in (0,1)$ such that $B_{e,\alpha}[x,r_0] \subseteq E$. So it is enough to show that $B_{e,\alpha}[u,\lambda r_0] \subseteq E$ for $\lambda r_0 \in (0,1)$.

Let $y \in B_{e,\alpha}[u,\lambda r_0]$ then $\nu_{(u-y),e}(\lambda r_0) \geq \alpha$.

Therefore, $\nu_{\lambda^{-1}(y-u),e}(r_0) = \nu_{(u-y),e}(\lambda r_0) \geq \alpha$ implies $x + \lambda^{-1}(y-u) \in B_{e,\alpha}[x,r_0]$.

Let $w = x + \lambda^{-1}(y-u)$ then $\lambda w = \lambda x + (y-u) \Rightarrow y = \lambda(w-x)+u$ $\Rightarrow y = \lambda(w-x) + \lambda x + (1-\lambda)a \Rightarrow y = (1-\lambda)a + w$ with $w, a \in E$.

Since E is convex, $y = (1-\lambda)a + w \in E$. Hence any point in (a, x) is an interior point of E. □

Corollary 11.2.1. *Let E be a convex subset of a 2-PN space (X, ν). Let x be an interior point of E and $y \in \overline{E}$ then $(x, y) \subseteq int(E)$.*

Proof. Suppose x is an interior point of E and $y \in \overline{E}$, then there exists a sequence $\{y_n\} \in E$ such that $y_n \longrightarrow y$. Let $z \in (x,y)$ then $z = \lambda x + (1-\lambda)y$ for some $\lambda \in (0,1)$. Define $z_n = \lambda x + (1-\lambda)y_n$. Since x is an interior point of E, then there exists $r_0 > 0, e \in X$ and $\alpha \in (0,1)$ such that $B_{e,\alpha}[x, r_0] \subseteq E$. By Theorem 11.2.3, $B_{e,\alpha}[z_n, \lambda r_0] \subseteq E$ for $\lambda r_0 \in (0,1)$ and $z_n \longrightarrow z$. Since ν is continuous for the first component and $z_n \longrightarrow z$ means that $\lim_{n\to\infty} \nu_{z_n - z, y}(t) = 1$ for $y \in X$ and $t > 0$. That is, there exists $n_0 \in \mathbf{N}$ such that $z_n \in B_{e,\alpha}[z_n, \lambda r_0]$ for every $n \geq n_0$. Now $\nu_{z_n - z, y}(t) = \nu_{z - z_n, y}(t)$ and we can say that $z \in B_{e,\alpha}[z_n, \lambda r_0] \subseteq E$. Hence $(x, y) \subseteq int(E)$. □

Corollary 11.2.2. *Let E be a non-empty convex subset of a 2-PN space (X, ν) then $\overline{int(E)} = \overline{E}$.*

Proof. It is obvious that $\overline{int(E)} \subseteq \overline{E}$. Let $y \in \overline{E}$ and take $x \in int(E)$. Then by Corollary 11.2.1, $(x, y) \subseteq int(E)$. If $\lambda_n \in (0,1)$ with $\lambda_n \to 0$ then $\{\lambda_n x + (1-\lambda_n)y\}$ is a sequence in (x, y).

Then $\nu_{[\lambda_n x + (1-\lambda_n)y] - y, z}(t) = \nu_{\lambda_n(x-y), z}(t) = N_{0,z}(t)$ for every $z \in X$, as $\lambda_n \to 0$. i.e. $\nu_{[\lambda_n x + (1-\lambda_n)y] - y, z}(t) = \varepsilon_0(t) \Rightarrow \nu_{[\lambda_n x + (1-\lambda_n)y] - y, z}(t) = 1$ for $t > 0$

$\Rightarrow \lambda_n x + (1-\lambda_n)y \longrightarrow y$ as $n \to \infty$. So, $y \in \overline{int(E)} \Rightarrow \overline{E} \subseteq \overline{int(E)}$. Hence $\overline{int(E)} = \overline{E}$. □

11.3 Compactness and Boundedness in 2-PN Spaces

Definition 11.3.1. A subset $E \subset X$ is said to be *compact* if each sequence of elements of X has a convergent subsequence in E.

Definition 11.3.2. Let F be a subset of a 2-PN space (X, ν). A convex series of elements of F is a series of the form $\Sigma_{n=1}^{\infty} \lambda_n x_n$ where $x_n \in F$ and $\lambda_n \geq 0$ for each n and $\Sigma_{n=1}^{\infty} \lambda_n = 1$.

The set F is said to be *convex series closed* if F contains the sum of every convergent convex series of its elements. Also, F is said to be *convex series compact* if every convex series of its elements is convergent to a point of F.

Remark 11.3.1. Every convex series compact set in a Menger's 2-PN space (X,ν) is convex series closed.

Proof. Let F be a convex series compact set in (X,ν), then there exists a convex series of elements of F, say $\Sigma_{n=1}^{\infty}\lambda_n x_n$ where $x_n \in F$ and $\lambda_n \geq 0$, which converges to some $x \in F$.

$\Rightarrow \lim_{n\to\infty} \nu_{\sum \lambda_n x_n - x, z}(t) = 1$ for all $z \in X$
$\Rightarrow \nu_{\sum \lambda_n x_n - x, z}(t) = \varepsilon_0(t)$ for all $z \in X$
$\Rightarrow \sum \lambda_n x_n - x$ and z are linearly dependent
$\Rightarrow \sum \lambda_n x_n - x = \lambda z$ for all $z \in X$

In particular for $\lambda z = z - x$, with $z \in F$ and $\sum \lambda_n x_n - x = (z-x)+x = z$. Hence F is convex series closed. □

Remark 11.3.2. Let F be a convex subset of a Menger's 2-PN space (X,ν) and $x_n \in F$ for $n \geq 1$. If $\sum \lambda_n = \lambda > 0$ where $\lambda_n \geq 0$ then $\sum \lambda^{-1}\lambda_n x_n$ is a convex series of elements of F. So, if $\sum \lambda_n x_n \to x$ then $x = \lambda a$ where $a \in \overline{F}$.

Proof. We have $\sum \lambda^{-1}\lambda_n x_n$ is a convex series of elements of F because $x_n \in F$ and $\lambda > 0$ with $\sum \lambda^{-1}\lambda_n = \lambda^{-1}\sum \lambda_n = \lambda^{-1}\lambda = 1$. Suppose $\sum \lambda_n x_n \to x$ then $\sum \lambda^{-1}\lambda_n x_n = \lambda^{-1}\sum \lambda_n x_n \to \lambda^{-1}x \in \overline{F}$, i.e.; $\lambda^{-1}x = a$ for some $a \in \overline{F}$ implies $x = \lambda a$. □

Theorem 11.3.1. *Let (X,ν) be a Menger's 2-PN space; then every closed convex subset of X is convex series closed.*

Proof. Let F be a closed convex subset of X and $\sum \lambda_n x_n$ be a convergent convex series of elements of F with sum x. We have $\sum \lambda_n x_n = x \Rightarrow x = \lambda_1 x_1 + \sum_{n=2}^{\infty} \lambda_n x_n$. Since $\sum_{n=1}^{\infty} \lambda_n = 1 \Rightarrow \sum_{n=2}^{\infty} \lambda_n = 1 - \lambda_1 > 0$. By Remark 11.3.2, $x = \lambda_1 x_1 + (1-\lambda_1)a$ where $a \in \overline{F}$ then $x \in F$. Hence F is convex series compact. □

Definition 11.3.3. A subset F of a Menger's 2-PN space (X,ν) is said to be *bounded* if for every $r \in (0,1)$ there exists $t_0 > 0$ such that $\nu_{x,y}(t_0) > 1-r$ for every $x \in F$ and $y \in X$.

Theorem 11.3.2. *A subset F of a 2-normed space $(X, \|\cdot,\cdot\|)$ is bounded if, and only if, F is bounded in the Menger's 2-PN space $(X, \frac{t}{t+\|\cdot,\cdot\|})$.*

Proof. Suppose that F is a bounded subset of $(X, \|\cdot,\cdot\|)$; then for every $x \in F$ there exists $M > 0$ such that $\|x, y\| \leq M$ for every $y \in X$. We have $\nu_{x,y}(t) = \frac{t}{t+\|x,y\|}$ for $x, y \in X$. Let $r \in (0, 1)$, choose $t_0 = \frac{M(1-r)}{r}$ then $t_0 > 0$ and $\nu_{x,y}(t_0) = \frac{t_0}{t_0+\|x,y\|} > \frac{t_0}{t_0+M} = 1 - r$. So, F is bounded in $(X, \frac{t}{t+\|\cdot,\cdot\|})$. Conversely, F is bounded in $(X, \frac{t}{t+\|\cdot,\cdot\|})$; then for every $r \in (0, 1)$ there exists $t_0 > 0$ such that $\nu_{x,y}(t_0) > 1 - r$ for every $x \in F$ and $y \in X$ implies $\frac{t_0}{t_0+\|x,y\|} > 1 - r$. Choose $M = \frac{t_0 r}{1-r}$ then $M > 0$ with $\|x, y\| < M$ for every $y \in X$. □

Theorem 11.3.3. *Let (X, ν) be a Menger's 2-PN space and F be a convex series compact subset of X; then*

(1) *F is convex series closed.*
(2) *F is bounded.*

The converse is true if X is complete.

Proof. (1) By Remark (2.3) it is clear.
(2) We prove this result by the contradiction method.

Let $r \in (0, 1)$ and $(a_n) \subset F$ such that $\nu_{a_n,z}(2^n) < 1 - r)$ for all n and $z \in X$. We have $\sum_{n=1}^{\infty} 2^{-n} = 1$ then $\sum_{n=1}^{\infty} 2^{-n} a_n$ is a convex series of elements of F. Since F is convex series compact, $\sum_{n=1}^{\infty} 2^{-n} a_n$ is convergent to some point in F. Hence $2^{-n} a_n$ converges to 0 as $n \to +\infty$ implies that for every $\epsilon > 0$ and $r \in (0, 1)$ there exists $k \in \mathbb{N}$ such that $\nu_{2^{-n} a_n, z}(t) > 1 - r$ for every $n \geq k$ and $t > 0$. In particular, $\nu_{2^{-n} a_n, z}(1) > 1 - r$ for every $n \geq k \Rightarrow \nu_{a_n,z}(2^n) > 1 - r$, a contradiction to our assumption. So, F is bounded.

Conversely, suppose that X is complete. Assume that (1) and (2) holds. One has to prove that F is convex series compact. Choose $r \in (0, 1)$. Since F is bounded there exists $t_0 > 0$ such that $\nu_{x,y}(t_0) > 1 - r$ for every $x \in F$ and $y \in X$. Let $\sum_{n=1}^{\infty} \lambda_n x_n$ be a convergent convex series of elements of F. If $\gamma_{n,m} = \sum_{i=n}^{m} \lambda_i$ then $\gamma_{n,m} \to 0$ as $n, m \to \infty$. Choose $t \in \mathbb{R}$; then there is $k \in \mathbb{N}$ such that $t\gamma_{n,m}^{-1} > 0$ for every $m, n \geq k$. Since F is bounded, $\nu_{x_n,z}(t\gamma_{n,m}^{-1}) > 1 - r$ implies

$$\begin{aligned}\nu_{\sum_{i=n}^{m} \lambda_i x_i, z}(t) &= \nu_{\sum_{i=n}^{m} \lambda_i x_i, z}(t\gamma_{n,m}^{-1}(\lambda_n + \lambda_{n+1} + \cdots + \lambda_m)) \\ &= \nu_{\sum_{i=n}^{m} \lambda_i x_i, z}(t\gamma_{n,m}^{-1}\lambda_n + t\gamma_{n,m}^{-1}\lambda_{n+1} + \cdots + t\gamma_{n,m}^{-1}\lambda_m)\end{aligned}$$

$$\geq \min\{\nu_{\lambda_n x_n, z}(t\gamma_{n,m}^{-1}\lambda_n), \nu_{\lambda_{n+1}x_{n+1},z}(t\gamma_{n,m}^{-1}\lambda_{n+1}), \ldots, \nu_{\lambda_m x_m, z}(t\gamma_{n,m}^{-1}\lambda_m)\}$$
$$= \min\{\nu_{x_n,z}(t\gamma_{n,m}^{-1}), \nu_{x_{n+1},z}(t\gamma_{n,m}^{-1}), \ldots, \nu_{x_m,z}(t\gamma_{n,m}^{-1})\}$$
$$> \min\{1-r, 1-r, \ldots, 1-r\}$$
$$= 1-r$$

i.e., $\sum_{n=1}^{\infty} \lambda_n x_n$ is a Cauchy sequence in X. So, $\sum_{n=1}^{\infty} \lambda_n x_n$ converges. Since F is convex series closed, the sum of $\sum_{n=1}^{\infty} \lambda_n x_n$ is in F. Hence F is convex series compact. □

Theorem 11.3.4. *Let (X, ν) be a Menger's 2-PN space and F be a complete, convex and bounded subset of X; then F is convex series compact.*

Proof. Suppose that $\sum_{n=1}^{\infty} \lambda_n x_n$ is a convex series of elements of F with $\lambda_n > 0$. By the same procedure in the above theorem $\sum_{n=1}^{\infty} \lambda_n x_n$ is a Cauchy sequence in X. Take $\alpha_n = \sum_{i=1}^{n} \lambda_i$ and $y_n = \sum_{i=1}^{n} \lambda_i x_i$. We show that $(\alpha^{-1} y_n)$ is a Cauchy sequence in F. Choose $r \in (0,1)$ and $t > 0$. Since F is bounded there exists $t_0 > 0$ such that $\nu_{x,y}(t_0) > 1-r$ for every $x \in F$ and $y \in X$. Let $z \in X$ then we have,

$$\nu_{y_n,z}(t_0) = \nu_{\sum_{i=1}^{n} \lambda_i x_i, z}(\alpha_n t_0))$$
$$\geq \min\{\nu_{\lambda_1 x_1,z}(\lambda_1 t_0), \nu_{\lambda_2 x_2,z}(\lambda_2 t_0), \ldots, \nu_{\lambda_n x_n,z}(\lambda_n t_0))\}$$
$$= \min\{\nu_{x_1,z}(t_0), \nu_{x_2,z}(t_0), \ldots, \nu_{x_n,z}(t_0)\}$$
$$> \min\{1-r, 1-r, \ldots, 1-r\}$$
$$= 1-r.$$

Since $\alpha_n^{-1} \to 1$ and (y_n) is a Cauchy sequence, there exists $k \in \mathbb{N}$ such that for all $z \in X$

$$\nu_{\alpha_n^{-1}y_n - \alpha_m^{-1}y_m, z}(t) = \nu_{\alpha_n^{-1}y_n - \alpha_m^{-1}y_n + \alpha_m^{-1}y_n - \alpha_m^{-1}y_m, z}\left(\frac{t}{2} + \frac{t}{2}\right)$$
$$\geq \min\left\{\nu_{\alpha_n^{-1}y_n - \alpha_m^{-1}y_n, z}\left(\frac{t}{2}\right), \nu_{\alpha_m^{-1}y_n - \alpha_m^{-1}y_m, z}\left(\frac{t}{2}\right)\right\}$$
$$= \min\left\{\nu_{y_n,z}\left(\frac{t}{2|\alpha_n^{-1} - \alpha_m^{-1}|}\right), \nu_{y_n - y_m, z}\left(\frac{\alpha_m t}{2}\right)\right\}$$
$$> \min\{1-r, 1-r, \ldots, 1-r\}$$
$$= 1-r$$

for every $n, m \geq k$.

Therefore, $(\alpha_n^{-1} y_n)$ is a Cauchy sequence in F and since F is complete, $(\alpha_n^{-1} y_n)$ converges to some $x \in F$. That is, there exists $k \in \mathbb{N}$ such that $\nu_{\alpha_n^{-1} y_n - x, z}(t) > 1 - r$ for every $z \in X$, and $n \geq k$ implies $\nu_{\alpha_n^{-1} \sum_{i=1}^n \lambda_i x_i - x, z}(t) > 1 - r$ for every $z \in X$, and $n \geq k$ as $n \to \infty$, we have $\nu_{\sum_{i=1}^n \lambda_i x_i - x, z}(t) = 1$ which implies $\lim_{n\to\infty} \nu_{y_n - x, z}(t) = 1$, which implies $y_n \to x$ and $x \in F$. Hence F is convex series compact. □

11.4 $\mathcal{D}$-Boundedness in 2-PN Spaces

Definition 11.4.1. Let A be a non-empty subset of a Menger's 2-PN space (X, ν) then its probabilistic radius R_A is defined by

$$R_A(x) := \begin{cases} l^- \varphi_A(x), & x \in [0, +\infty), \\ 1, & x = \infty. \end{cases}$$

where $\varphi_A(x) := \inf\{\nu_{x,y}(x) : x, y \in A\}$.

Definition 11.4.2. Let A be a non-empty subset of a Menger's 2-PN space (X, ν) then A is said to be:

(1) *certainly bounded*, if $R_A(x_0) = 1$ for some $x_0 \in (0, \infty)$;
(2) *perhaps bounded*, if one has $R_A(x) < 1$ for every $x \in (0, \infty)$ and $l^{-1} R_A(+\infty) = 1$;
(3) *perhaps unbounded*, if $R_A(x_0) > 0$ for some $x_0 \in (0, \infty)$ and $l^{-1} R_A(+\infty) \in (0, 1)$;
(4) *certainly unbounded*, if $l^{-1} R_A(+\infty) = 0$.

A is said to be $\mathcal{D}$-bounded if either (1) or (2) holds.

Theorem 11.4.1. *Let (X, ν) be a Menger's 2-PN space. If $|\alpha| \leq |\beta|$ then $\nu_{\beta x, y}(t) \leq \nu_{\alpha x, y}(t)$ for every $x, y \in X$ and $\alpha, \beta \in \mathbb{R} - \{0\}$.*

Proof. We have $\nu_{\beta x, y}(t) = \nu_{x,y}(\frac{t}{|\beta|})$ and $\nu_{\alpha x, y}(t) = \nu_{x,y}(\frac{t}{|\alpha|})$. Since $|\alpha| \leq |\beta|$ then $\frac{t}{|\beta|} \leq \frac{t}{|\alpha|} \Rightarrow \nu_{x,y}(\frac{t}{|\beta|}) \leq \nu_{x,y}(\frac{t}{|\alpha|}) \Rightarrow \nu_{\beta x, y}(t) \leq \nu_{\alpha x, y}(t)$. □

Theorem 11.4.2. *Let (X, ν) and A be a Menger's 2-PN space and a non-empty subset respectively, then A is $\mathcal{D}$-bounded if, and only if, there exists a d.f. $G \in \mathcal{D}^+$ such that $\nu_{x,y} \geq G$ for every $x, y \in A$.*

Proof. Suppose that A is $\mathcal{D}$-bounded, then there exists $R_A \in \mathcal{D}^+$. Choose $G := R_A$, then $\nu_{x,y} \geq G$ for every $x, y \in A$. Conversely, suppose that there is a d.f. $G \in \mathcal{D}^+$ such that $\nu_{x,y} \geq G$ for every $x, y \in A$. It implies

$\ell^- \inf_{x,y\in A} \nu_{x,y}(t) \geq \inf G(t) \Rightarrow R_A(t) \geq G(t) \Rightarrow \lim_{t\to\infty} R_A(t) \geq \lim_{t\to\infty} G(t) = 1$. So, A is $\mathcal{D}$-bounded. □

We denote the set of all $\mathcal{D}$-bounded subsets in a Menger's generalized 2-probabilistic normed space $(X \times Y, \nu)$ (MG2PN space) by $\mathcal{P}_{\mathcal{D}^+}(X \times Y)$.

Theorem 11.4.3. *Let $(X \times Y, \nu)$ and $A \times B$, $C \times B$ be a Menger's G2PN space and two non-empty $\mathcal{D}$-bounded subsets of $X \times Y$ respectively. Then $(A + C) \times B$ is a $\mathcal{D}$-bounded set if $\mathcal{D}^+$ is a closed set under the t-norm M, i.e. $M(\mathcal{D}^+ \times \mathcal{D}^+) \subseteq \mathcal{D}^+$.*

Proof. For every $(a, b) \in A \times B$ and $(c, b) \in C \times B$ one has $(a + c, b) \in (A + C) \times B$. Therefore

$$\nu_{a+c,b} \geq M\{\nu_{a,b}, \nu_{c,b}\} \geq M\{\nu_{a,b}, R_{C\times B}\} \geq M\{R_{A\times B}, R_{C\times B}\},$$

and as a consequence

$$R_{(A+C)\times B} \geq M\{R_{A\times B}, R_{C\times B}\}.$$

According to the hypothesis we have

$$M\{R_{A\times B}, R_{C\times B}\} \subseteq \mathcal{D}^+,$$

and finally $\ell^- R_{(A+C)\times B}(+\infty) = 1$. □

Theorem 11.4.4. *Let $(X \times Y, \nu)$ and $A \times B$, $C \times D$, $A \times D$, $C \times B$ be a Menger's G2PN space and four non-empty $\mathcal{D}$-bounded subsets of $X \times Y$ respectively. Then the set given by*

$$A \times B + C \times D := \{(p, q) + (r, s) = (p + r, q + s)\}$$

is $\mathcal{D}$-bounded if $\mathcal{D}^+$ is a closed set under the t-norm M.

Proof. By (MG2PN-3) one has, for all $(p, q) \in A \times B$, $(r, s) \in C \times D$,

$$\begin{aligned}\nu_{(p,q)+(r,s)} &\geq M\{\nu_{p,q+s}, \nu_{r,q+s}\}\\ &\geq M\{R_{A\times(B+D)}, \nu_{r,q+s}\} \geq M\{R_{A\times(B+D)}, R_{C\times(B+D)}\},\end{aligned}$$

and as a consequence

$$R_{A\times B+C\times D} \geq M\{R_{A\times(B+D)}, R_{C\times(B+D)}\}.$$ □

Bibliography

J. Aczél. (1966), *Lectures on Functional Equations and their Applications*, Academic Press, New York.

C. Alsina, B. Schweizer. (1983), *The countable products of probabilistic metric spaces*, Houston J. Math. **9**, pp.303–310.

C. Alsina, M.J. Frank, B. Schweizer. (2006), *Associative Functions, Triangular Norms and Copulas*, World Scientific, Singapore.

C. Alsina. (1983), *On convex triangle functions*, Aequationes Math., **26**, pp.191–196.

C. Alsina, B. Schweizer. (1981), *On a theorem of Mouchtari and Šerstnev*, Note di Mathematica, **Vol. I**, pp.19–24.

C. Alsina. (1987), On stability of a functional equation arising in probabilistic normed spaces, in *General Inequalities 5, 5th International Conference on General Inequalities*, Oberwolfach, May 4–10, 1986, Birkhäuser Verlag, Basel, pp.263–271.

C. Alsina, B. Schweizer, A. Sklar. (1997), *Continuity properties of probabilistic norms*, J. Math. Anal. Appl., **208**, pp.446–452.

C. Alsina, B. Schweizer, A. Sklar. (1993), *On the definition of a probabilistic normed space*, Aequationes Math., **46**, pp.91–98.

C. Alsina, B. Schweizer, C. Sempi, A. Sklar. (1997), *On the definition of a probabilistic inner product space*, Rend. Mat., **(7)17**, pp.115–127.

C. Alsina, M.J. Frank., B. Schweizer. (2005), *Associative Functions on Intervals: A primer of triangular norms*, World Scientific, Hackensack.

H. Bauer. (1996), *Probability Theory*, de Gruyter, Berlin-New York.

P. Billinsgley. (1979), *Probability and Measure*, 3rd edition, Wiley, New York.

S. Chang, Y. J. Cho, S. M. Kang. (2001), *Nonlinear operator theory in probabilistic metric spaces*, Nova Science Publishers, Inc, New York.

L. Cădariu and V. Radu. (2004), *On the stability of the Cauchy functional equation: A fixed point approach*, in: J. Sousa Ramos, D. Gronau, C. Mira, L. Reich, A.N. Sharkovsky (Eds.), Iteration Theory ECIT 02, in: Gracer Math. Ber., **346**, pp.323–350.

G. Constantin, I. Istratescu. (1989), *Elements of Probabilistic Analysis*, Kluwer Academic Publishers, Brooks/Cole, Pacific Grove CA.

D. Dugué. (1955), *L'existence d'une norme est incompatible avec la convergence en probabilité*, C. R. Acad. Sci. Paris, **240**, pp.1037–1039.

D. Dugué. (1956), *Incompatibilité de la convergence presque certaine et de l'écart*, C. R. Acad. Sci. Paris, **242**, pp.728–729.

N. Dunford, J.T. Schwartz. (1957), *Linear operators*, Part I: General theory, Wiley, New York.

R.E. Edwards. (1995), *Functional Analysis: Theory and Applications*, Dover, New York.

G.Z. Eskandani. (2008), *On the stability of an additive functional equation in quasi-Banach spaces*, J. Math. Anal. Appl. **345**, pp.405-409.

K. Fan. (1944), *Entferung zweier zufälliger Griössen und die Konvergenz nach Wahrschein-lichkeit*, Math Z., **49**, pp.681–683.

C. Felbin. (1992), *Finite dimensional fuzzy normed linear space*, Fuzzy Sets and Systems, **48**, no. 2, pp.239–248.

X. Fernique. (1998), *Un modèle presque sûr pour la convergence en loi*, C. R. Acad. Sci. Paris Sér. I., **306**, pp.335–338.

J.M. Fortuny. (1984), *Espacios con producto interior probabilístico*, Stochastica, **8**, pp.229–248.

M.J. Frank. (1971), *Probabilistic topological spaces*, J. Math. Anal. Appl., **34**, pp.67–81.

M.J. Frank. (1975), *Associativity in a class of operations on a space of distribution functions*, Aequationes Math., **12**, pp.121–144.

M.J. Frank, B. Schweizer. (1979), *On the duality of generalized infimal and supremal convolutions*, Rend. Mat., **(6)12**, pp.1–23.

M. Fréchet. (1950), *Géneralités sur los probabilitiés. Eléments aleátoires*, Gauthier-Villars, Paris.

R. Fritsche. (1971), *Topologies for probabilistic metric spaces*, Fund. Math. **72**, pp.7–16.

Z. Gajda. (1991), *On the stability of additive mapping*, Int. J. Math. Math. Sci., **14**, pp.431–434.

M.B. Ghaemi, B. Lafuerza-Guillén, S. Saiedinezhad. (2009), *Invariant and semi-invariant probabilistic normed spaces*, Chaos, Solitons and Fractals (Elsevier), **42**, pp.256–264.

I. Golet. (2007), *Approximation theorems in probabilistic normed spaces*, Novi Sad J. Math., **38**, no.3, pp.73–79.

O. Hadžić, and E. Pap. (2001), *Fixed point theory in probabilistic metric spaces*, Kluwer Academic Publishers, Dordrecht.

T.L. Hicks, P.L. Sharma. (1984), *Probabilistic metric structures. Topological classification*, Review of Research Faculty of Science, Univ. of Novi Sad, **14**, pp.43–50.

U. Höhle. (1977), *Probabilistic metrization of generalized topologies*, Bull. Acad. Polon. Sci. **25**, pp.493–498.

P.K. Harikrishnan, B. Lafuerza-Guillen, K.T. Ravindran. (2011), *Accretive operators and Banach Alogolu theorem in linear 2-normed spaces*, Proyeccions Journal of Mathematics, **30**, no.3, pp.319–327.

P.K. Harikrishnan. (2013), *Some Aspects of Accretive and other Operators in Abstract Spaces*, Ph.D Thesis, Kannur University, India.

I.M. James. (1990), *Introduction to Uniform Spaces*, Cambridge University Press, Cambridge.

J. Jacod, P. Protter. (2000), *Probability Essentials*, Springer, Berlin-Heidelberg.

W. Jarczyk, J. Matkowski. (2002), *On Mulholland's inequality*, Proc. Amer. Math. Soc. **130**, pp.3243–3247.

I. H. Jebril, R. Ibrahim, M. Ali. (2002), *Bounded linear operators in probabilistic normed spaces*, J. Inequalities Pure Applied Math., **4**, no.1, article 8.

P. Jordan, J. Von Neumann. (1935), *On inner products in linear metric spaces*, Ann. of Math. **36**, pp.719–723.

C.F.K. Jung. (1969), *On generalized complete metric spaces*, Bull. Amer. Math. Soc. **75**, pp.113–116.

J.L. Kelley. (1955, 1975), *General topology*, Van Nostrand, New York; reprinted, Graduate Texts in Mathematics **27**, Springer, New York-Heidelberg-Berlin.

E.P. Klement, R. Mesiar, Eds. (2005), *Logical, Algebraic, Analytic, and Probabilistic Aspects of Triangular Norms*, Elsevier, Amsterdam.

E.P. Klement, R. Mesiar, E. Pap. (2000), *Triangular Norms*, Kluwer, Dordrecht.

A.N. Kolmogorov. (1934, 1991), *Zur normierbarkeit eines allgemeinen topologischen linearen Räumes*, Studia Math., **5**, pp.29–33; English translation in V.M. Tikhomirov (Ed.), *Selected Works of A.N. Kolmogorov, Vol. I: Mathematics and Mechanics*, Kluwer, Dordrecht-Boston-London, pp.183–186.

M.S. Krasnoselškii, Y.B. Rutickii. (1961), *Convex Function and Orlicz Spaces*, Noordhof, Groningen.

F. Lael, K. Nourouzi. (2009), *Compact operators defined on 2-normed and 2-probabilistic normed spaces*, Hindawi Publishing Corporation, Mathematical Problems in Engineering, Volume 2009, Article ID 950234, 17 pages.

B. Lafuerza-Guillén. (1996), *Primeros Resultados en el estudio de los espacios normados probabilisticos con nuevos conceptos de acotacin*, Ph.D Thesis, Universidad de Almeria, Spain.

B. Lafuerza-Guillén. (2004), *Finite products of probabilistic normed spaces*, Radovi Matematički, **13**, pp.111–117.

B. Lafuerza-Guillén, A. Rodríguez Lallena, C. Sempi. (1999), *A study of boundedness in probabilistic normed spaces*, J. Math. Anal. Appl., **232**, pp.183–196.

B. Lafuerza-Guillén. (2001), *$\mathcal{D}$-bounded sets in probabilistic normed spaces and their products*, Rend. Mat., Serie VII, **21**, pp.17–28.

B. Lafuerza-Guillén, C. Sempi, G. Zhang. (2010), *A study of boundedness in probabilistic normed spaces*, Nonlinear Analysis, **73**, no.5, pp.1127–1135.

B. Lafuerza-Guillén, J.A. Rodríguez Lallena, C. Sempi. (2008), *Normability of probabilistic normed spaces*, Note di Matematica, **29**, no.1, pp.99–111.

B. Lafuerza-Guillén, J.A. Rodríguez Lallena, C. Sempi. (1998), *Probabilistic norms for linear operators*, J. Math. Anal. Appl., **220**, pp.462–476.

B. Lafuerza-Guillén, J.A. Rodríguez Lallena, C. Sempi. (1997), *Some classes of probabilistic normed spaces*. Rend. Mat., **17**, pp.237–252.

B. Lafuerza-Guillén, J.L. Rodríguez. (2008), *Translation invarient generalised topologies induced by probabilistic norms*, Note de Mathematica, **29**, no.1, pp.157–164.

B. Lafuerza-Guillén, J.A. Rodríguez Lallena, C. Sempi. (1995), *Completion of probabilistic normed spaces*, Intrnat. J. Math. and Math. Sci., **18**, pp.649–652.

P. Lévy. (1937), *Distance de deux variables aléatoires et distance de deux lois de probabilité, Calcul des Probabilités et ses Applications*, **1**, pp.331–337.

P. Lévy. (1973–1980),*Œuvres de Paul Lévy*, Gauthier-Villars, Paris.

E. Lukacs. (1975a) *Stochastic Convergence*, Academic Press, New York-San Francisco-London.

E. Marczewski. (1955), *Remarks on the convergence of measurable sets and measurable functions*, Coll. Math. **3**, pp.118–124; Reproduced in E. Marczewski. (1996), *Collected mathematical papers*, Polish Academy of Sciences, Warsaw, pp.475–481.

K. Menger. (1942), *Statistical metrics*, Proc. Nat. Acad. Sci., USA, **28**, pp.535–537.

D. Mihet, V. Radu. (2007), *Generalized pseudo-metrices and fixed points in probabilistic metric spaces*, Carpathian J. Math., **23**(1–2), pp.126–132.

D. Mihet, V. Radu. (2008), *On the stability of additive Cauchy functional equation in random normed spaces*, J. Math. Anal. Appl., **343**, pp.567–572.

D. Mihet, R. Saadati, S.M. Vaezpour. (2010), *The stability of the quatric functional equation in random normed spaces*, Acta. Appl. Math., **110**, pp.797–803.

D. Mihet, R. Saadati, S.M. Vaezpour. (2011), *The stability of an additive functional equation in Menger probabilistic ϕ-normed spaces*, Mathematica Slovaca, **61(5)**, pp.817–826.

A.K. Mirmostafee, M.S. Moslehian. (2008), *Fuzzy version of Hyers-Ulam-Rassias theorem*, Fuzzy sets and systems, **159**, pp.720–729.

R. Moynihan, B. Schweizer, A. Sklar. (1978), *Inequalities among operations on probability distribution functions*, in E.F. Beckenbach (Ed), *General Inequalities*, Birkhauser, Basel, pp.133–149.

B. Morrel, J. Nagata. (1978), *Statistical metric spaces as related to topological spaces*, General Top. Appl., **9**, pp.233–237.

D.H. Mushtari, A.N. Šerstnev. (1966), *On methods of introducing a topology in random metric spaces*, Izv. Vyssh. Uch. Zav. Math. **6(55)**, pp.99–106.

D.H. Mushtari. (1967), *The completion of random metric spaces*, Kazan Gos. Univ. Učen. Zap., **167**, no.3, pp.109–119.

R.B. Nelsen. (2006), *An Introduction to Copulas*, Springer, New York, 2nd ed.

A. Pourmoslemi, M. Salimi. (2008), *$\mathcal{D}$-bounded sets in generalized probabilistic 2-normed spaces*, World Applied Sciences Journal **3(2)**, pp.265–268, ISSN 1818-4952.

B.J. Prochaska. (1967), *On Random Normed Spaces*, Ph.D Dissertation, Clemson University.

V. Radu. (1975), *Linear operators in random normed spaces*, Bull. Math. Soc. Sci. Math. R.S. Roumanie N.S., **17**, pp.217–220.

V. Radu. (1975), *Sur une norme aléatoire et la continuité des opérateurs linéaires dans les espaces normés aléatoires*, C.R. Acad. Paris, **280A**, pp.1303–1305.

V. Radu. (2003), *The fixed point alternative and the stability of functional equations*, Sen. Fixed Point Theory, **4(1)**, pp.91–96.

M.M. Rao. (1987), *Measure Theory and Integration*, Wiley, New York.

M.M. Rao, Z.D. Ren. (1991), *Theory of Orlicz Spaces*, Marcel Dekker, New York.

R. Saadati, S.M. Vaezpour. (2008), *Linear operators in probabilistic normed spaces*, J. Math. Anal. Appl., **346**, no.2, pp.446–450.

R. Saadati, M. Amini. (2005), *D-boundedness and D-compactness in finite dimensional probabilistic normed spaces*, Proc. Indian Acad. Sci. Math. Sci., **115**, no.4, pp.483–492.

R. Saadati, G.A. Zhang, B. Lafuerza-Guillén. (2012), *Total boundedness in probabilistic normed spaces*, U.P.B. Sci. Bull. Series A, **74**, no.2, pp.1–10.

S. Saminger-Platz, C. Sempi. (2008), *A primer on triangle functions-I*, Aequationes Math., **76**, pp.201–240.

S. Saminger-Platz, C. Sempi. (2010), *A primer on triangle functions-II*, Aequationes Math., **80**, pp.239–268.

P. Sarkoci., *Dominance is not transitive on continuous triangular norms*, Aequationes Math (in press).

H.H. Schäfer., *Topological Vector Spaces*, Springer-Verlag, Heidelberg.

B. Schweizer, A. Sklar. (1983, 2005), *Probabilistic Metric Spaces 2nd edition*, Springer, North Holland.

B. Schweizer. (1975), *Multiplication on the space of probability distribution functions*, Aequationes Math. **12**, pp.153–183.

B. Schweizer. (1991), Thirty years of copulas, in G. DallÁglio, S. Kotz, G. Salinetti (Eds.), *Advances in probability distributions with n given marginals; beyond the copulas, Mathematics and its Applications*, Kluwer, Dordrecht, pp.13–50.

B. Schweizer. (2003), *Commentary on probabilistic geometry*, Selecta Mathematica (Springer), pp.409–432.

B. Schweizer. (2005), *Triangular Norms, Looking Back Triangle Functions Looking Ahead*, (English) Klement, Erich Peter (Ed.) *et al.*, pp.3–15.

B. Schweizer, A. Sklar. (1958), *Espaces métriques aléatoires*, C.R. Acad. Sci. Paris, **247**, pp.2092–2094.

B. Schweizer, A. Sklar. (1960), *Statistical metric spaces*, Pacific J. Math. **10**, pp.313–334.

B. Schweizer, A. Sklar. (1962), *Statistical metric spaces arising from sets of random variable in Euclidean n spaces*, Theory Probab. Appl., **7**, pp.447–456.

B. Schweizer, A. Sklar. (1963), *Triangle inequalities in the class of statistical metric spaces*, J. London. Math. Soc., **38**, pp.401–406.

B. Schweizer, A. Sklar. (1973), *Probabilistic metric spaces determined by measure-preserving transformations*, Z. Warsch. Verw. Gebiete, **26**, pp.235–239.

B. Schweizer, A. Sklar. (1974), *Operations on distribution functions not derivable from operations on random variables*, Studia Math. **52**, pp.43–52.

B. Schweizer, A. Sklar. (1980), *How to derive all L_p-metrics from a single probabilistic metric*, General Inequalities, Proceedings of the Second

International Conference on General Inequalities, E.F. Beckenbach (Ed), Birkhauser-Basel (Boston-Stuttgart), pp.429–434.

C. Sempi. (1982), *On the space of distribution functions*, Riv. Mat. Uni. Parma, **(4)8**, pp.243–250.

C. Sempi. (1985), *Orlicz metrics derive from a single probabilistic metric*, Stochastica, **9**, pp.181–184.

C. Sempi. (1992), *Hausdorff distance and the completion of probabilistic normed spaces*, Bull. Un. Mat. Ital., **(7)6-B**, pp.317–327.

C. Sempi. (2004), *Probabilistic metric spaces*, Encyclopedia of General Topology, K.P Hart, J. Nagata, J.E. Vaughn (Eds). Kluwer, Dordrecht.

C. Sempi. (2006), *A short and partial history of probabilistic normed spaces*, Mediterr. J. Math., **3**, pp.283–300.

M.L. Senchal. (1965), *Approximate Functional Equations and Probabilistic Inner Product Spaces*, Illinois Institute of Technology, Ph.D. Thesis.

A.N. Šerstnev. (1962), *Random normed spaces: problems of completeness*, Kazan Gos. Univ. Učhen Zap., **122**, pp.3–20.

A.N. Šerstnev. (1963), *On the notion of a random normed space*, Dold. Akad. Nauk SSSR, **149(2)**, pp.280–283 (English Translation in Soviet Math. Dokl., **4**, pp.388–390).

A.N. Šerstnev. (1963), *Best approximation problem in random normed spaces*, Dokl. Acad. Nauk. SSSR, **149**, no.3, pp.539–542.

A.N. Šerstnev. (1964), *On a probabilistic generalization of a metric spaces*, Kazan Gos. Univ. Učen. Zap., **124**, pp.3–11.

A.N. Šerstnev. (1964), *Some best approximation problems in random normed spaces*, Rev. Roumaine Math. Pures Appl., **9**, pp.771–789.

H. Sherwood. (1966), *On the completion of probabilistic metric spaces*, Z. Wahrsch. Verw. Gebiete, **6**, pp. 62–64.

H. Sherwood. (1971), *Complete probabilistic metric spaces*, Z. Wahrsch. Verw. Gebiete, **20**, pp.117–128.

H. Sherwood. (1969), *On E-spaces and their relation to other classes of probabilistic metric spaces*, J. London Math. Soc., **44**, pp.441–448.

H. Sherwood. (1979), *Isomorphically isometric probabilistic normed linear spaces*, Stochastica, **3**, pp.71–77.

H. Sherwood. (1984), *Characterizing dominates in a familiy of triangular norms*, Aequationes Math., **27**, pp.255–273.

D.A. Sibley. (1971), *A metric for weak convergence of distribution functions*, Rocky Mountain J. Math., **1**, pp.427–430.

A.V. Skorohod. (1965, 1982), *Studies in the Theory of Random Processes*, Addison-Wesley, Reading; reprinted by Dover, New York.

R.M. Tardiff. (1976), *Topologies for probabilistic metric spaces*, Pacific J. Math. **65**, pp.233–251.

R.M. Tardiff. (1984), *On a generalized Minkowski inequality and its relation to dominates for t-norms*, Aequationes Math., **27**, pp.308–316.

M.D. Taylor. (1985), *Introduction to Functional Analysis*, Wiley, New York-London-Sydney.

M.D. Taylor. (1985), *New metrics for weak convergence of distribution functions*, Stochastica, **9**, pp.5–17.

A.J. Thomasian. (1956), *Distances et normes sur les espaces de variables aléatoires*, C. R. Acad. Sci. Paris, **242**, pp.447–448.

A.J. Thomasian. (1956), *Metrics and norms for spaces of random variables*, Ann. Math. Statist., **28**, pp.512–514.

E.O. Thorp. (1962), *Generalized topologies for statistical metric spaces*, Fund. Math., **51**, pp.9–21.

A. Wald. (1943, 1955), *On a statistical generalization of metric spaces*, Proc. Nat. Acad. Sci. USA, **29**, pp.196–197; reprinted in Abraham Wald, *Selected Papers in Statistics and Probability*, Wiley, New York.

A. Wilansky. (1964), *Functional Analysis*, Blaisdell Publishing Co. (Ginn and Co.), New York-Toronto-London.

D. Williams. (1991), *Probability with Martingales*, Cambridge University Press, Cambridge.

C.X. Wu, M. Ming. (1990), *Fuzzy norms, Probabilistic norms and Fuzzy metrics*, Fuzzy Information Processing, Fuzzy Sets and Systems, **36**, pp.137–144.

G. Zhang, M. Zhang. (2008), *On the normability of generalised Šerstnev PN spaces*, J. Math. Anal. Appl., **340**, pp.1000–1011.

Index

Printed in the United States
By Bookmasters